Making Hole

Unit II, Lesson 1
Third Edition

by William E. Jackson

Published by

PETROLEUM EXTENSION SERVICE
The University of Texas at Austin
Continuing Education
Austin, Texas

in cooperation with

INTERNATIONAL ASSOCIATION
OF DRILLING CONTRACTORS
Houston, Texas

2000

Library of Congress Cataloging-in-Publication Data

Jackson, William E., 1929—
    Making hole / by William E. Jackson. — 3rd ed.
        p. cm. — (Rotary drilling series ; unit II, lesson 1)
        ISBN 0-88698-190-5 (alk. paper)
    1. Oil well drilling. 2. Oil well drilling rigs. I. Title. II. Series.
TN871.2.J318   2000
622'.3381—dc21   00-024964
                                                                    CIP

Brand names, company names, trademarks, or other identifying symbols appearing in illustrations and/or text are used for educational purposes only and do not constitute an endorsement by the author or the publisher.

Catalog no. 2.20130
ISBN 0-88698-190-5

No state tax funds were used to publish this book.
The University of Texas at Austin is an equal opportunity employer.

Contents

▼
▼
▼

Figures

Tables

▼
▼
▼

Foreword

Drilling an oil or gas well, or "making hole," as people in the oil industry sometimes refer to it, is a complicated operation. Drilling situations vary a great deal throughout the world. Consequently, it is difficult to explain the fine points of field geology, engineering, technology, and economics that apply to every well. Moreover, in spite of great advances in drilling technology, as much art as science is sometimes involved in drilling a well. For a particular well, even experienced drillers and engineers may not agree on the right way to make hole. Nevertheless, basic principles apply to all wells and this book emphasizes these factors.

Thus, this third edition of *Making Hole* deals in a general way with the factors that affect the rate of penetration in drilling. It includes the basics of well planning and cost control, and takes a brief look at some recent innovations in drilling technology.

Keep in mind that this manual is for training purposes only; readers should therefore be aware that nothing in it is approval or disapproval of any product or practice. PETEX made every effort to ensure accuracy, but on occasion, mistakes may occur. Indeed, when readers find errors, please inform us so that we can correct future reprints and editions. Moreover, PETEX welcomes suggestions from readers that would make this book better. Although PETEX received a great deal of assistance from many manufacturers, suppliers, and contractors, PETEX is wholly responsible for the book's content.

Ron Baker

Acknowledgments

▼
▼
▼

The author expresses a sincere appreciation to the many people who contributed to this edition of *Making Hole*. Those who provided illustrations, background information and/or also reviewed manuscript drafts include:

John Baer, Division Engineer, Helmerich & Payne International Drilling Co., Oklahoma City, OK.

Randy Brown, District Engineer, Hughes Christensen, Oklahoma City, OK.

Mark Franklin, Sales Engineer and Brian Jeffery, Field Engineer, Reed-Hycalog Tool Company, Oklahoma City, OK.

R. L. Hilbun, Licensed Professional Engineer and owner, Summa Engineering Inc., Oklahoma city, OK.

Ken Fischer and Jason McFarland with IADC, Houston, TX, provided many official IADC forms and charts for inclusion in this edition.

Others who kindly provided illustrations and permissions to publish their material include:

Tom Rogers, Mid-Continent Area Manager, Baker Hughes INTEC, Oklahoma City, OK.

Tom Enegonio, Sperry-Sun Drilling Services, Oklahoma City, OK.

The staff at the Oklahoma Commission on Marginally Producing Wells in Norman, OK, provided generous use of materials from their excellent library.

My sincere thanks to all.

Units of Measurement

Throughout the world, two systems of measurement dominate: the English system and the metric system. Today, the United States is almost the only country that employs the English system.

The English system uses the pound as the unit of weight, the foot as the unit of length, and the gallon as the unit of capacity. In the English system, for example, 1 foot equals 12 inches, 1 yard equals 36 inches, and 1 mile equals 5,280 feet or 1,760 yards.

The metric system uses the gram as the unit of weight, the metre as the unit of length, and the litre as the unit of capacity. In the metric system, for example, 1 metre equals 10 decimetres, 100 centimetres, or 1,000 millimetres. A kilometre equals 1,000 metres. The metric system, unlike the English system, uses a base of 10; thus, it is easy to convert from one unit to another. To convert from one unit to another in the English system, you must memorize or look up the values.

In the late 1970s, the Eleventh General Conference on Weights and Measures described and adopted the Système International (SI) d'Unités. Conference participants based the SI system on the metric system and designed it as an international standard of measurement.

The *Rotary Drilling Series* gives both English and SI units. And because the SI system employs the British spelling of many of the terms, the book follows those spelling rules as well. The unit of length, for example, is *metre*, not *meter*. (Note, however, that the unit of weight is *gram*, not *gramme*.)

To aid U.S. readers in making and understanding the conversion to the SI system, we include the following table.

English-to-SI-Units Conversion Factors

Quantity or Property	English Units	Multiply English Units By	To Obtain These SI Units
Length, depth, or height	inches (in.)	25.4	millimetres (mm)
		2.54	centimetres (cm)
	feet (ft)	0.3048	metres (m)
	yards (yd)	0.9144	metres (m)
	miles (mi)	1609.344	metres (m)
		1.61	kilometres (km)
Hole and pipe diameters, bit size	inches (in.)	25.4	millimetres (mm)
Drilling rate	feet per hour (ft/h)	0.3048	metres per hour (m/h)
Weight on bit	pounds (lb)	0.445	decanewtons (dN)
Nozzle size	32nds of an inch	0.8	millimetres (mm)
Volume	barrels (bbl)	0.159	cubic metres (m³)
		159	litres (L)
	gallons per stroke (gal/stroke)	0.00379	cubic metres per stroke (m³/stroke)
	ounces (oz)	29.57	millilitres (mL)
	cubic inches (in.³)	16.387	cubic centimetres (cm³)
	cubic feet (ft³)	28.3169	litres (L)
		0.0283	cubic metres (m³)
	quarts (qt)	0.9464	litres (L)
	gallons (gal)	3.7854	litres (L)
	gallons (gal)	0.00379	cubic metres (m³)
	pounds per barrel (lb/bbl)	2.895	kilograms per cubic metre (kg/m³)
	barrels per ton (bbl/tn)	0.175	cubic metres per tonne (m³/t)
Pump output and flow rate	gallons per minute (gpm)	0.00379	cubic metres per minute (m³/min)
	gallons per hour (gph)	0.00379	cubic metres per hour (m³/h)
	barrels per stroke (bbl/stroke)	0.159	cubic metres per stroke (m³/stroke)
	barrels per minute (bbl/min)	0.159	cubic metres per minute (m³/min)
Pressure	pounds per square inch (psi)	6.895	kilopascals (kPa)
		0.006895	megapascals (MPa)
Temperature	degrees Fahrenheit (°F)	$\dfrac{°F - 32}{1.8}$	degrees Celsius (°C)
Thermal gradient	1°F per 60 feet	—	1°C per 33 metres
Mass (weight)	ounces (oz)	28.35	grams (g)
	pounds (lb)	453.59	grams (g)
		0.4536	kilograms (kg)
	tons (tn)	0.9072	tonnes (t)
	pounds per foot (lb/ft)	1.488	kilograms per metre (kg/m)
Mud weight	pounds per gallon (ppg)	119.82	kilograms per cubic metre (kg/m³)
	pounds per cubic foot (lb/ft³)	16.0	kilograms per cubic metre (kg/m³)
Pressure gradient	pounds per square inch per foot (psi/ft)	22.621	kilopascals per metre (kPa/m)
Funnel viscosity	seconds per quart (s/qt)	1.057	seconds per litre (s/L)
Yield point	pounds per 100 square feet (lb/100 ft²)	0.48	pascals (Pa)
Gel strength	pounds per 100 square feet (lb/100 ft²)	0.48	pascals (Pa)
Filter cake thickness	32nds of an inch	0.8	millimetres (mm)
Power	horsepower (hp)	0.75	kilowatts (kW)
Area	square inches (in.²)	6.45	square centimetres (cm²)
	square feet (ft²)	0.0929	square metres (m²)
	square yards (yd²)	0.8361	square metres (m²)
	square miles (mi²)	2.59	square kilometres (km²)
	acre (ac)	0.40	hectare (ha)
Drilling line wear	ton-miles (tn•mi)	14.317	megajoules (MJ)
		1.459	tonne-kilometres (t•km)
Torque	foot-pounds (ft•lb)	1.3558	newton metres (N•m)

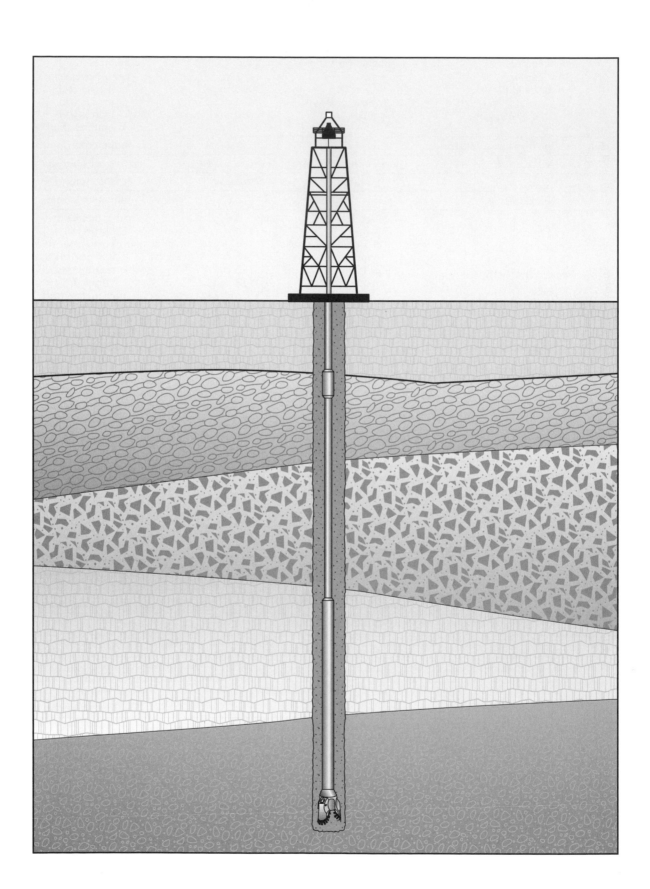

Introduction

It's true: to drill a hole you put the bit on the bottom and turn it to the right. In simple terms, that's how you "make hole"; but, of course, the whole story is more complicated. What type of bit do you put on bottom? How much weight do you put on the bit? Do you rotate the bit fast or slow? What about mud properties? What should pump pressure be? All these questions, and more, are related and critical to drilling progress. Thus, if one factor changes, it can result in unforeseen difficulties unless the crew makes other adjustments as drilling proceeds.

Drilling situations vary widely throughout the world. A successful drilling program in South Louisiana could be wrong for a contractor in Oklahoma's hard-rock country. A well plan for drilling the deep overpressured gas zones of West Texas could not be used in California's shallow tar sands. A wildcat well only a few miles from a producing field can encounter vastly different conditions. The field may be in flat-lying beds, for instance, while the wildcat may encounter steeply dipping beds. Further, not all wells are drilled vertically. The operator may specify that a deviated or a horizontal hole be kicked off at a certain depth.

To safely operate in such widely divergent conditions, every drilling operation must be carefully planned. Whatever the conditions, the drilling contractor's goal is the same: to drill a usable hole to the operator's specifications for the least possible cost. Indeed, the contractor's survival depends on meeting that goal. The contractor must usually accomplish the objectives set out in the drilling contract (fig. 1) in the shortest time possible. (A "footage contract" means that the operator pays the contractor so much money for every foot (ft) or metre (m) of well the contractor drills, regardless of how long it takes. "Day rate" means that the operator pays the contractor so much money for every day the contractor is drilling, regardless of how many feet per day the rig drills.)

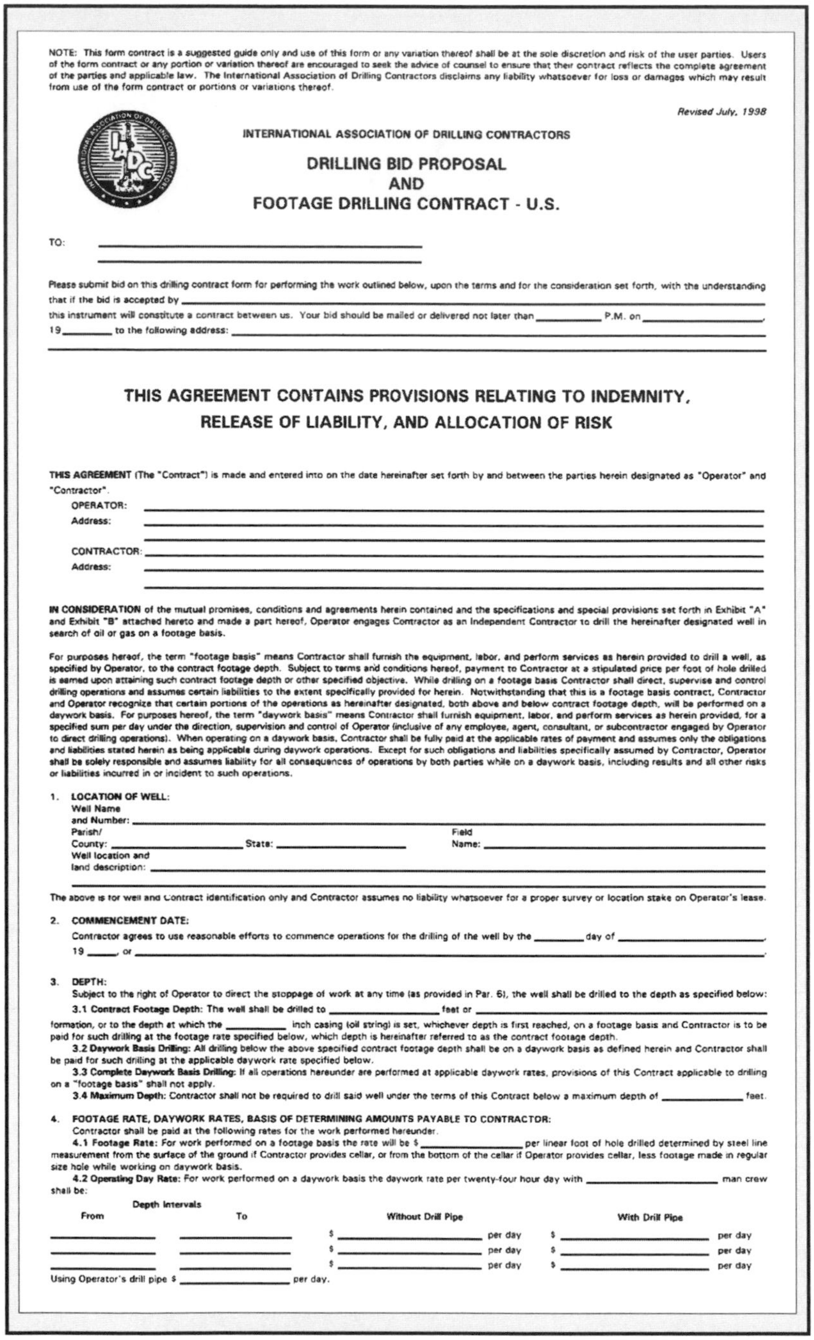

NOTE: This form contract is a suggested guide only and use of this form or any variation thereof shall be at the sole discretion and risk of the user parties. Users of the form contract or any portion or variation thereof are encouraged to seek the advice of counsel to ensure that their contract reflects the complete agreement of the parties and applicable law. The International Association of Drilling Contractors disclaims any liability whatsoever for loss or damages which may result from use of the form contract or portions or variations thereof.

Revised July, 1998

INTERNATIONAL ASSOCIATION OF DRILLING CONTRACTORS

DRILLING BID PROPOSAL
AND
FOOTAGE DRILLING CONTRACT - U.S.

TO: ____________________
____

Please submit bid on this drilling contract form for performing the work outlined below, upon the terms and for the consideration set forth, with the understanding that if the bid is accepted by ____________________ this instrument will constitute a contract between us. Your bid should be mailed or delivered not later than ________ P.M. on ____________, 19______ to the following address: ____________________

THIS AGREEMENT CONTAINS PROVISIONS RELATING TO INDEMNITY, RELEASE OF LIABILITY, AND ALLOCATION OF RISK

THIS AGREEMENT (The "Contract") is made and entered into on the date hereinafter set forth by and between the parties herein designated as "Operator" and "Contractor".

OPERATOR: ____________________
Address: ____________________

CONTRACTOR: ____________________
Address: ____________________

IN CONSIDERATION of the mutual promises, conditions and agreements herein contained and the specifications and special provisions set forth in Exhibit "A" and Exhibit "B" attached hereto and made a part hereof, Operator engages Contractor as an Independent Contractor to drill the hereinafter designated well in search of oil or gas on a footage basis.

For purposes hereof, the term "footage basis" means Contractor shall furnish the equipment, labor, and perform services as herein provided to drill a well, as specified by Operator, to the contract footage depth. Subject to terms and conditions hereof, payment to Contractor at a stipulated price per foot of hole drilled is earned upon attaining such contract footage depth or other specified objective. While drilling on a footage basis Contractor shall direct, supervise and control drilling operations and assumes certain liabilities to the extent specifically provided for herein. Notwithstanding that this is a footage basis contract, Contractor and Operator recognize that certain portions of the operations as hereinafter designated, both above and below contract footage depth, will be performed on a daywork basis. For purposes hereof, the term "daywork basis" means Contractor shall furnish equipment, labor, and perform services as herein provided, for a specified sum per day under the direction, supervision and control of Operator (inclusive of any employee, agent, consultant, or subcontractor engaged by Operator to direct drilling operations). When operating on a daywork basis, Contractor shall be fully paid at the applicable rates of payment and assumes only the obligations and liabilities stated herein as being applicable during daywork operations. Except for such obligations and liabilities specifically assumed by Contractor, Operator shall be solely responsible and assumes liability for all consequences of operations by both parties while on a daywork basis, including results and all other risks or liabilities incurred in or incident to such operations.

1. **LOCATION OF WELL:**
Well Name and Number: ____________________
Parish/County: ____________ State: ____________ Field Name: ____________
Well location and land description: ____________________

The above is for well and Contract identification only and Contractor assumes no liability whatsoever for a proper survey or location stake on Operator's lease.

2. **COMMENCEMENT DATE:**
Contractor agrees to use reasonable efforts to commence operations for the drilling of the well by the ______ day of ____________, 19 ____, or ____________________.

3. **DEPTH:**
Subject to the right of Operator to direct the stoppage of work at any time (as provided in Par. 6), the well shall be drilled to the depth as specified below:
3.1 Contract Footage Depth: The well shall be drilled to ____________ feet or ____________ formation, or to the depth at which the ________ inch casing (oil string) is set, whichever depth is first reached, on a footage basis and Contractor is to be paid for such drilling at the footage rate specified below, which depth is hereinafter referred to as the contract footage depth.
3.2 Daywork Basis Drilling: All drilling below the above specified contract footage depth shall be on a daywork basis as defined herein and Contractor shall be paid for such drilling at the applicable daywork rate specified below.
3.3 Complete Daywork Basis Drilling: If all operations hereunder are performed at applicable daywork rates, provisions of this Contract applicable to drilling on a "footage basis" shall not apply.
3.4 Maximum Depth: Contractor shall not be required to drill said well under the terms of this Contract below a maximum depth of ________ feet.

4. **FOOTAGE RATE, DAYWORK RATES, BASIS OF DETERMINING AMOUNTS PAYABLE TO CONTRACTOR:**
Contractor shall be paid at the following rates for the work performed hereunder.
4.1 Footage Rate: For work performed on a footage basis the rate will be $ ____________ per linear foot of hole drilled determined by steel line measurement from the surface of the ground if Contractor provides cellar, or from the bottom of the cellar if Operator provides cellar, less footage made in regular size hole while working on daywork basis.
4.2 Operating Day Rate: For work performed on a daywork basis the daywork rate per twenty-four hour day with ____________ man crew shall be:

Depth Intervals From	To	Without Drill Pipe	With Drill Pipe
________	________	$ ________ per day	$ ________ per day
________	________	$ ________ per day	$ ________ per day
________	________	$ ________ per day	$ ________ per day

Using Operator's drill pipe $ ____________ per day.

Figure 1. IADC footage drilling bid-contract (Courtesy of IADC)

Cost per foot (metre) drilled is the ultimate gauge of drilling success. A hole drilled fast, but crooked or out of gauge, can create costly problems that wipe out the apparent savings. Similarly, a hole drilled in picture-perfect condition can have such high mud costs or take so many bits and so much rig time that the venture becomes a business failure. The following equation can be used to calculate the cost per foot (metre) drilled:

$$C_t = \frac{B + C_r\,(t + T)}{F}$$

where

C_t = cost per foot or metre drilled
B = bit cost
C_r = rig cost in dollars per hour
t = rotation time in hours
T = tripping time in hours
F = footage (metreage) per bit.

For example, suppose that a bit cost $10,000 and that rig costs are $400 per hour. Further, the total time a bit was on bottom and rotating was 128 hours, crew members spent 10 hours tripping, and the bit drilled 1,920 feet (585 metres). Thus,

$$C_t = \frac{10,000 + 400\,(128 + 10)}{1,920}$$

$$= \frac{10,000 + 400\,(138)}{1,920}$$

$$= \frac{10,000 + 55,200}{1,920}$$

$$= \frac{65,200}{1,920}$$

$$C_t = \text{about } \$34/\text{ft } (\$112/\text{m}).$$

In general, contractors can reduce drilling costs by maximizing rotating time and minimizing tripping time. Although rig costs vary from rig to rig and bit costs change from well to well, correct drilling practices can assure that each bit drills further, which reduces tripping time. A driller is concerned with six basic factors that affect the rate of penetration (ROP):

1. bit selection;
2. weight on bit;
3. rotary speed;
4. drilling fluid properties;
5. hydraulics; and
6. formation properties.

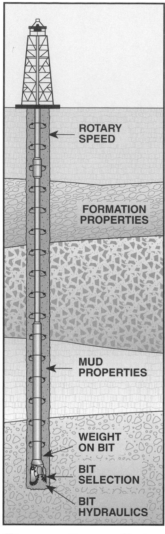

A successful drilling operation requires the most efficient combination of the six factors for the formation being drilled (fig. 2). These basic factors remain the same for a vertical, deviated, or horizontal hole.

To summarize—
- The goal of any drilling project is to drill a usable hole to the operator's specifications for the least possible cost.
- Cost per foot (metre) of hole gauges drilling success; the equation to calculate cost per foot (metre) of hole is:

$$C_t = \frac{B + C_r\,(t + T)}{F}$$

Factors that affect ROP
- bit selection
- weight on bit
- rotary speed
- drilling fluid properties
- hydraulics
- formation properties

Figure 2. Factors affecting penetration rate

Well Planning

Good well planning is the first step towards successful, least-cost, or "optimized drilling." Hill and Lee in an article in the oilfield magazine *World Oil*, defined optimized drilling as, "the well plan and drilling operation that results in the lowest-cost well meeting final construction specifications and drilled safely with respect to people, property, and the environment."

Every well can be planned with the benefit of information from previous drilling in the area, whether the planned well is an offset well or a remote wildcat. Examples of information available include bit records, mud reports, electrical logs, mud logs, daily drilling reports, and geological reports. The computer allows the well planner to try numerous combinations of complex factors to arrive at a final well plan. The best planning, using the best and most complete information, still does not replace the learning curve (fig. 3), which demonstrates that the experience gained in early high-cost wells permits savings on subsequent, similar wells.

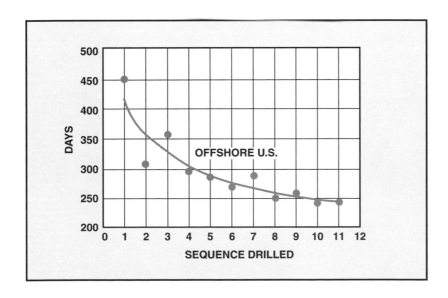

Figure 3. "The learning curve" for drilling offshore development wells (Courtesy of World Oil)

Selecting the right rig to drill the hole is a critical part of well planning. API Bulletin D10, *Procedure for Selecting Rotary Drilling Equipment*, offers a method of coordinating rig specifications with the demands of the drilling operation. The rig's specifications must fit the conditions of the drilling operation. The rig must be properly rated for depth, hoisting capacity, pumps, and other criteria that affect the rig's ability to do the job. For example, in soft-rock drilling, the rig's circulating system must have good hole-cleaning ability; in hard-rock drilling, on the other hand, the rig must have the hoisting capacity to handle a heavier drill string. A rig's circulation system is a critical factor when downhole motors or turbines are used. Compromises may be necessary, but a rig can always be found and adapted to handle the drilling conditions expected.

Before drilling begins, a pre-spud meeting is usually held at which representatives of the operator and the contractor review the drilling contract. These meetings benefit those directly involved in drilling the well. Among the items discussed are the location preparation, time of move-in, water supply, vendor and service contractors, mud program, well data reporting, the well program, including well testing, coring, deviation, geological hazards, environmental precautions, or any special provisions included in the contract. The pre-spud meeting thus establishes a plan of action and sets out how certain problems will be handled should they arise.

It is impossible to foresee everything that may occur during drilling; therefore, the rig manager (toolpusher) and the operator's representative must make many, possibly critical, decisions while drilling is underway. The rig manager and the drilling crew have the responsibility for rig maintenance, safety, service company performance, and overall execution of the well program unless operations fall under the day rate provisions of the drilling contract.

To summarize—

- Well planning is the operation that results in the lowest-cost well meeting final construction specifications and drilled safely with respect to people, property, and the environment.
- To plan new wells, information from previously drilled wells used includes—
    - bit records
    - mud reports
    - electric logs
    - mud logs
    - drilling reports
    - geologic reports
- Earliest wells drilled usually cost more than later wells in the same field.
- Rig specifications must meet the demands of the drilling situation.
- Pre-spud meetings are essential to reduce later drilling costs.

Bits

The ideal bit is always the one that does the job for the least overall cost, but the great variety of available bits complicates the bit selection process. Manufacturers make bits for virtually every drilling need. Given reasonable time, bit manufacturers can deliver custom-designed bits for any given drilling situation. With the aid of computers, a bit supplier can review all previous drilling records in an area and deliver a customized, recommended bit program to customers (fig. 4).

A good geological prognosis is an invaluable aid in selecting the proper bit. The prognosis provides critical data on the formations and type rock to be expected at depth. For instance, if soft shale and medium limestone are expected from 6,800 to 7,400 ft (2,072 to 2,256 m) and hard, cherty limestone from 7,400 to 7,750 ft (2,256 to 2,362 m), the bit program can be set up accordingly.

Bit Selection

* FIGURE OUT IF YOU CAN FIND A BIT THAT WILL MAKE IT ALL IN ONE TRIP.

- *OFFSET DATA*
- *BIT REP*
- *PROVEN TRACK RECORD*

HUGHES CHRISTENSEN — RECOMMENDED BIT PROGRAM

GEOGRAPHIC LOCATION	OPERATOR	CONTRACTOR / RIG	PREPARED FOR
Gulfo de Paria Oeste	Conoco Venezuela	Mallard 76	E. J. GILBERT

FIELD / AREA	LOCATION / WELL NUMBER	SALES REPRESENTATIVE	DATE
	Esmeralda	Raul Salazar	AUGUST 7, 1998

No.	SIZE (in.)	BIT TYPE	DEPTH OUT (ft)	DIST DRLD (ft)	DRLG TIME (hrs)	ROP (ft/hr)	ACC TIME (hrs)	WOB (klb)	RPM	MUD WT (ppg)	DAYS	REMARKS
1	26	GTX-CG1	2000	2000	25.0	80.0	25.0	2 15	100 160		1.2	GT TECHNOLOGY
			Set 20 in. casing. 1.5 Days.								2.7	
2	16	GTX-CG1	4800	2800	21.0	133.3	46.0	5 30	100 160		3.8	STH / CENTER JET
3	16	GTX-CG1	7300	2500	19.5	128.2	65.5	5 30	100 180		4.7	GT TECHNOLOGY
4	16	GTX-CG1	9000	1700	14.0	121.4	79.5	5 30	100 200		5.4	
			Set 13-3/8 in. casing. 2 Days.								7.4	
5	12-1/4	GTX-C1	9100	100	5.0	20.0	84.5	5 30	100 160		7.6	DRILL OUT FLOAT EQUIP.
6	14	14RWD512	12400	3300	100.0	33.0	185	4 12	60 180		14.5	RWD OPENS HOLE TO 14"
7	8-1/2	R526U4	12401	1	1.0	1.0	186	4 12	60 180		16.6	3/4' PDC CUTTERS
			Set 11-3/4 in. casing. 2 Days.								18.6	
8	10-5/8	GT-1	12500	99	5.0	19.8	191	5 30	100 160		18.8	DRILL OUT FLOAT EQUIP.
9	12-1/4	12.25RWD51	14500	2000	90.0	22.2	281	4 12	60 180		23.0	RWD OPENS HOLE TO 12.25
10	8-1/2	R526U4	14501	1	1.0	1.0	282	4 12	60 180		25.1	3/4' PDC CUTTERS RERUN BIT
			Set 9-5/8 in. casing. 2 Days.								27.1	
11	8-1/2	R526U4	18500	3999	170.0	23.5	452	5 20	60 180		33.9	3/4' PDC CUTTERS

Figure 4. A recommended bit or well program (Courtesy of Hughes Christensen)

That is, a bit designed for drilling soft-to-medium hard formations would be selected for the first formation and a bit designed for drilling hard formations for the second formation. Or, it may be possible to select a bit that could drill both types of formation and thereby save a trip to change bits.

Where it is not possible to predict formation tops or rock characteristics, a computer system tied to sophisticated downhole instruments can provide information while drilling. Such systems are termed "measurement while drilling" (MWD) and can identify the rock type, hardness, compressive strength, porosity, and other properties—information that can influence bit selection and trip scheduling (fig. 5).

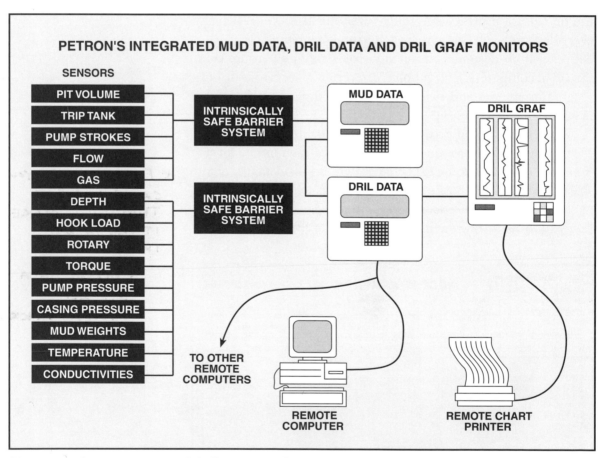

Figure 5. A rig site integrated drilling system for measurement-while-drilling (MWD)

Bit Design

The most important consideration affecting bit design is the type of rock the bit will be drilling. Is the formation hard or soft? Is it composed of abrasive sand? Is it a sticky, heavy shale? Is it porous chalk? In nature, subsurface formations change with depth. Unfortunately, however, no single bit can drill these variable rock types with equal efficiency. If that were the case, rotary drilling would be a simple, automated process from surface to total depth. The right bit must be selected to drill different formations (fig. 6). Knowing how bits drill and how they are designed, as well as being aware of formation (rock) properties, can facilitate the bit selection process.

All drilling bits are of two basic designs—roller cone bits, also called "rock bits," and fixed cutter bits, sometimes referred to as "fixed head," or "shear bits." Roller cone bits are either steel tooth (milled or forged), or tungsten carbide inserts. Two types of fixed cutter bits are available: natural diamond or synthetic diamond.

Figure 6. Bits

How Bits Drill

No matter the type, all bits are designed to crush, scrape, plow, or shear a rock with a substance that is harder than the rock, namely steel, tungsten carbide alloy, or a natural or synthetic diamond (fig. 7). A roller cone bit crushes the rock because of the great weight applied on the rock by the bit's cutters. By offsetting the cones, the crushing action is combined with a scraping action that is highly effective, especially in soft formations.

A natural diamond bit plows the rock by pushing the rock aside to form a groove, much like a plowed furrow, grinding the rock like a millstone. This type bit drills very slowly but is effective for hard, abrasive formations and in smaller size bits.

A synthetic diamond (polycrystalline diamond compact or PDC) shears or slices the rock and, like natural diamond bits, is usually used on fixed cutter bits. Shearing the rock is an efficient drilling action that takes only a third of the weight and energy required to crush rock.

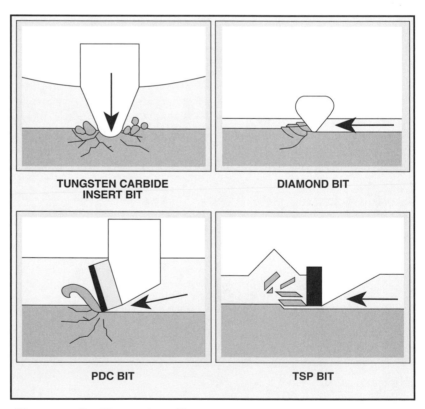

Figure 7. Drilling action of bits

Roller cone bits are made with cones mounted on rugged bearings. The bearings may be made of balls, rollers, or journals, or a combination of the three types (fig. 8). Journal bearings distribute the load over a maximum area and therefore tolerate the highest weight on the bit. The cone rolls around the bottom of the hole as the drill string is rotated or as the bit is turned by a downhole motor. There are usually three cones (tricone bits) but some designs may have two or only one.

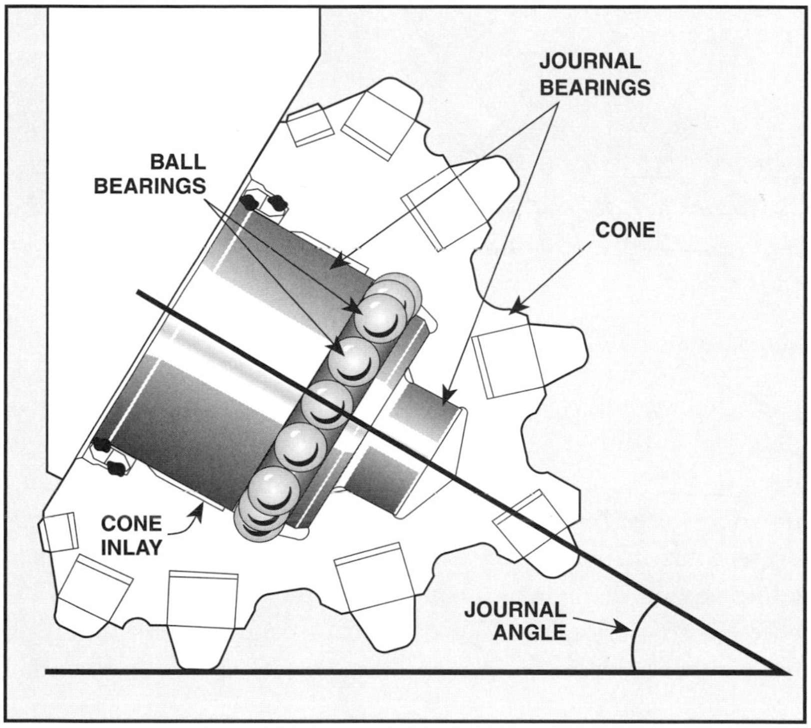

Roller Cone Bits

Figure 8.   Combination journal and ball bearings in a roller cone bit showing the journal angle

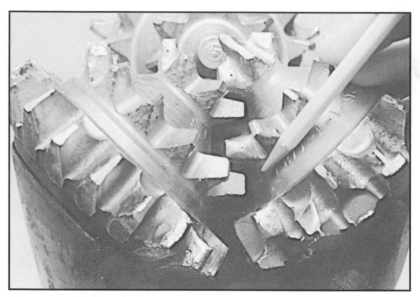

Figure 9.   The rows of cutters on one cone intermesh with the cutters on the other cones.

Each cone has rows of teeth that do the actual cutting. The teeth (cutters) may be either steel, tungsten carbide inserts, or diamond enhanced inserts. Steel-tooth bits have teeth that have been milled out of the cone body or, less often, forged from a metal powder. Insert bits have teeth made of tungsten carbide that are pressed into drilled openings in the cone. The placement, number, composition, shape, and quality of the teeth or inserts affect the efficiency of the bit. The rows of teeth are carefully interfit to allow the rows of teeth on one cone to project into spaces on the other cones as the bit rotates (fig. 9).

Steel-Tooth Bits

Steel-tooth bits are used for a wide variety of formations. Table 1 shows how bit design varies with the formation drilled. Generally, soft formation bits have fewer and longer teeth and hard formation bits have more, but shorter, teeth.

Manufacturers install the cones of soft-formation bits with offset, which means the cones do not rotate around the center of the bit (fig. 10). Offset imparts a twisting, scraping action that is highly effective in soft formations. Soft formations include gypsum, red beds, clays, marl, and soft shale. Soft-formation bits have widely spaced, self-cleaning teeth (fig. 11). This design prevents interference from the cuttings that cause a bit to ball up. Soft-formation

a. On-center

b. Off-center

Figure 10.   Cone offset. Each cone rotates on its own axis.

Table 1
Steel Tooth Bit Specifications Relative to Formation Type

Formation	Quantity of Teeth	Tooth Spacing	Tooth Height	Tooth Sharpness Angle	Journal Angle Degrees	Offset Angle Degrees	Hardmetal
Soft IADC 1XX	Few	Wide	Tall	39-42	32-33	2-5	Partial to full
Medium IADC 2XX	Moderate	Moderate	Medium	43-45	32-36	2-3	None to full
Hard IADC 3XX	Many	Close	Short	45-50	33-36	0-2	None to partial

Courtesy of Reed-Hycalog

Figure 11. Soft-formation milled-tooth bit (Courtesy of Hughes Christensen)

bits have thinner cones and smaller, lighter bearings and journals; therefore, they are best used with lighter weights on the bit and with higher rotary speed (rpm).

In soft limestone and tough, waxy shale, or in soft formations containing thin beds of hard rock, a slight variation in bit specifications may result in faster penetration rates. For instance, the same offset might be retained but more teeth per cone and tungsten carbide hardfacing could reduce tooth abrasion and improve the footage per bit.

Dolomite and hard limestone require higher weights on the bit to overcome the greater compressive strength of these harder rocks. Typically, bits used to drill harder formations have more, but shorter, teeth and have less cone offset. These changes in bit design allow heavier cones and bearings, which hold up under the added weight required, to achieve the crushing action needed to drill the harder rocks. Tooth hardfacing also may enhance and extend bit life in these formations.

The hardest rocks are quartzite, granite, dolomite, and hard limestone containing chert and quartz sand. The compressive strength of these rocks may vary but the most efficient steel-tooth bit will have closely spaced, short teeth, no cone offset, and high-strength bearings. Thus constructed, the heavy bit weight needed to crush the rock can be effectively employed. Insert bits have now virtually replaced tooth bits for drilling the hardest rocks, because tungsten carbide is considerably harder than the steel used in tooth bits and is therefore usually more effective in hard, abrasive rocks.

Insert Bits   Insert bits have chisel, conical, or hemispherical tungsten carbide insert cutters (compacts) set in the cones instead of steel teeth (fig. 12). Insert bits were originally developed to drill the hardest rock, which quickly wore out conventional steel-tooth bits. However, they were generally unsatisfactory in soft rocks or formations with variable hard and soft streaks. Because the cones were relatively soft, cone erosion was a serious problem in the early models but improvements in metallurgy have resulted in modern insert bits that are extremely durable and able to drill in many different formations (figs. 13, 14, and 15). In some areas, like the mid-continental U.S., they are the dominant bit type. The decision to change to an insert bit is based on economics. Insert bits cost twice as much as steel-tooth bits. However, they provide more consistent penetration rates and last longer, which means more drilling hours and fewer trips.

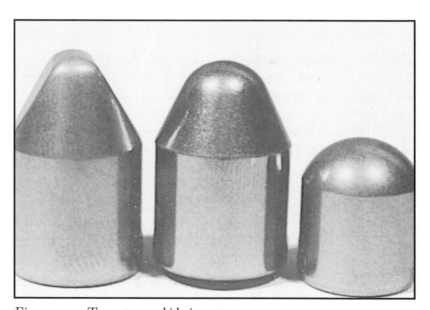

Figure 12.   Tungsten carbide inserts

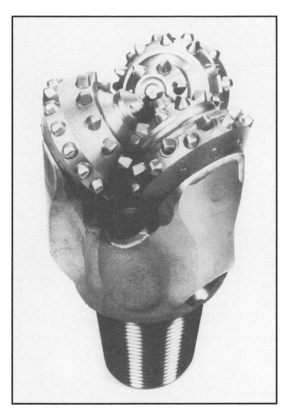

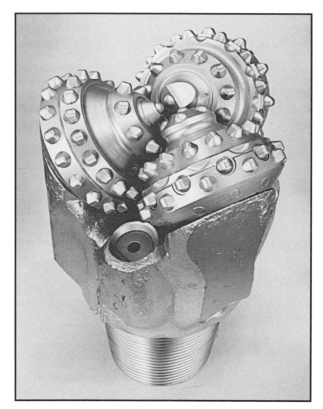

Figure 13. Soft-formation insert bit (Courtesy of Reed-Hycalog)

Figure 14. Medium-formation insert bit (Courtesy of Hughes Christensen)

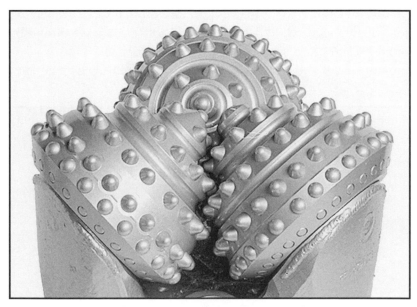

Figure 15. Hard-formation insert bit (Courtesy of Hughes Christensen)

Table 2
Insert Bit Specifications Relative to Formation Type

Formation	Quantity of Teeth	Tooth Spacing	Tooth Height	Journal Angle Degrees	Offset Angle Degrees
Soft IADC 4XX-5XX	Few	Wide	Tall	32-33	2-5
Medium IADC 6XX	Moderate	Moderate	Medium	32-36	0-3
Hard IADC 7XX-8XX	Many	Close	Short	36	0-2

Courtesy of Reed-Hycalog

Table 2 is a guide to insert bit design for various formations. Note that the shape, offset, and placement of the inserts vary in the same way as steel-tooth bits and generally affect drilling efficiency in a similar manner. Tungsten carbide inserts in the outer area of the bit (fig. 16) help prevent wear that results in an undergauge hole. Hardfacing, tungsten carbide inserts, or both hardfacing and inserts on the exterior of the bit's body also help extend the drilling life of the bit.

Natural diamond bits and PDC bits, which are also called fixed-cutter bits, are used when the performance of roller cone bits falls below some rate (feet or metres per hour—usually established in a long shaley section). When such conditions occur, a diamond bit or a PDC bit offers a good alternative. These types of bits have no moving parts and are thus radically different from cone bits.

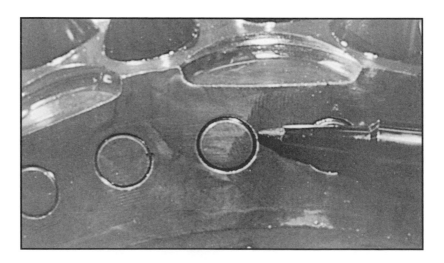

Figure 16. Flat tungsten carbide gauge row inserts

Because there are no moving parts, the general term "fixed-head bits" is sometimes used. They have cutting surfaces of either natural diamond, synthetic diamond, or diamond impregnated material. As always, cost must be considered before a decision is made. Natural diamond bits, polycrystalline diamond compact (PDC), or diamond impregnated bits can cost up to $30,000, several times the cost of an insert bit. Bit suppliers can offer alternatives to buying these expensive bits. They can also be leased or rented, usually on a specified charge per foot drilled.

In spite of the higher price, a natural diamond bit or a PDC bit can often be a cost-effective choice. In some situations, like deviated or horizontal holes drilled with downhole motors, low weight, or high rotation speeds, they may be mechanically necessary.

Natural Diamond Bits

The natural diamonds used in drilling bits are industrial-grade stones of various size and shape. The body of the bit is made of a tungsten carbide powder and a bonding agent that results in a metal that is harder than steel. A diamond bit has a one-piece, cone-shaped body with a center fluid passage (fig. 17). The bit's ability to drill a particular formation is determined by the bit shape and the size, number, and placement of the diamonds.

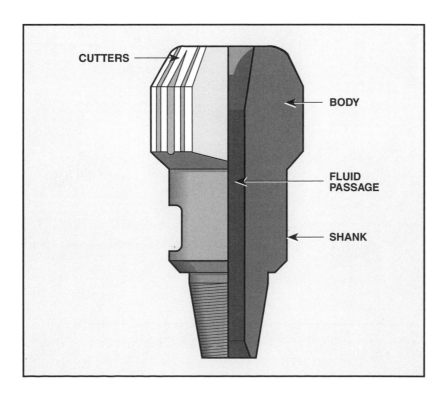

Figure 17. Four parts of a diamond bit

There are no nozzles. The pattern of the watercourses directs the flow of drilling fluid across the bit-rock interface (fig. 18). This is an important design factor because diamonds are thermally unstable; if overheated they decompose. Proper cooling by the drilling fluid is therefore critical and circulation must be carefully watched.

Diamond bits are nearly always used on rocks classified as hard. Like cone bits, the size of the diamond cutters varies by the formation drilled—fewer, larger diamonds for drilling softer rock and smaller, densely-set diamonds for harder rock. The shape of the bit body (profile) also varies for different type formations. In general, steeper-side cones are used in softer formations and the flatter profile bits are used in harder rocks. When diamond bits are used, the hole must be totally junk free to avoid breaking the diamonds.

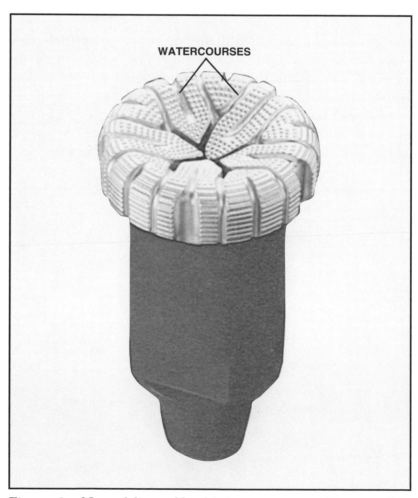

Figure 18. Natural diamond bit showing watercourses (Courtesy of Reed-Hycalog)

The original rotary bit, the fishtail bit, (named for its design) was a *drag bit*. Modern drag bits, now called *fixed-cutter bits*, were introduced in 1976 and are rapidly becoming as common as roller cone bits. These polycrystalline diamond compact (PDC) bits have cutters made of synthetic diamond crystals bonded to a tungsten carbide insert brazed into pockets in the body or blades of the bit (fig. 19). Like a natural diamond bit, a PDC bit has no moving parts. The cutters sit sideways in clusters on the blades of the bit (figs. 20, 21).

Fixed-Cutter Bits

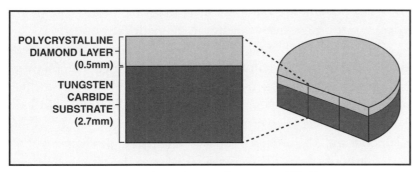

Figure 19. A polycrystalline diamond compact (PDC)

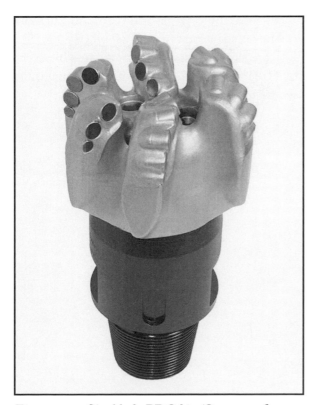

Figure 20. Six-blade PDC bit (Courtesy of Hughes Christensen)

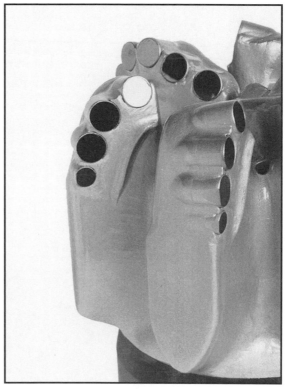

Figure 21. Detail of PDCs set in blade (Courtesy of Hughes Christensen)

Figure 22. Combination TSP and natural diamond bit (Courtesy of Reed-Hycalog)

The fixed cutters scrape the rock away as the drill string and bit rotate together (see fig. 7) or as the bit is turned by a downhole motor or turbine. Nozzles direct the flow of drilling fluid and cuttings similar to roller cone bits. The combination of a diamond's hardness and abrasion resistance with tungsten carbide's durability allows long bit runs. PDC bits require less weight to scrape and shear the rock so they are efficient when run at high rotation speeds. This is especially important in horizontal drilling applications. PDC bits are expensive, but, when compared to conventional bits, their fast penetration rates and increased time in the hole (fewer trips) result in significant savings in many drilling situations.

The main disadvantage of the PDC bit is that the synthetic diamond is less stable than a natural diamond at high temperatures. Synthetic diamonds called TSPs (thermally stable polycrystalline) are more stable at higher temperatures than the PDC. TSPs are larger and closer in size to a natural diamond than to a PDC diamond and are usually round or triangular in shape. A TSP functions directly as a cutter and is cast into the bit matrix rather than being brazed into pockets like a PDC insert. Both types are mounted sideways on the bit and exert a similar shearing action. The TSP bit profile also resembles a natural diamond bit more than a PDC (fig. 22). The penetration rate of TSP bits is about like a diamond bit and both are slower than PDCs or tricone bits. The main disadvantage of TSPs is their poor performance in shale, and that has limited their use to date.

Hybrid Bits

Hybrid bits attempt to combine the best qualities of natural diamonds, PDCs, TSPs, or tungsten carbide insert cutters in one bit. For example, some PDC bits will use natural diamonds or TSPs as gauge cutters where their slow rate of penetration is unimportant because they are reaming and not actually cutting new hole. Other combinations include a diamond-impregnated pad in the gauge area of a PDC bit (fig. 23) or a shorter diamond-impregnated backup stud behind a PDC cutter (fig. 24). As the bit wears, the backup stud assumes part of the load (and the heat), thus extending the life of the PDC. This type of hybrid works well in soft formations that become harder at depth or that have stringers of hard rock.

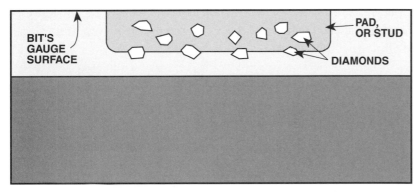

Figure 23. Diamond-impregnated pad set in gauge surface of bit body

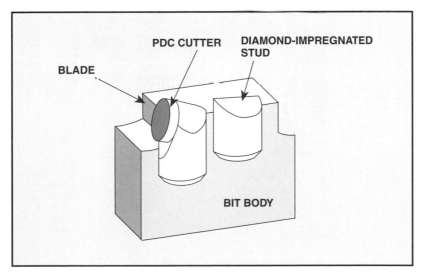

Figure 24. Diamond-impregnated stud backup to a PDC cutter

Special Purpose Bits

Some bits are designed to accomplish a specific task, such as cutting a core or starting a sidetracked hole. Because these bits are only used for special situations, the decision to run them is not usually within the routine bit selection process. Often, their use is preplanned when drilling operations reach a certain depth or formation.

Special purpose bits may be either roller cone or fixed-head designs. The more common cone bits include those with extended nozzles, two-cone bits, air bits, and jet-deflection bits (a type of two-cone bit). Ream-while-drilling (RWD) bits are designed to enlarge a hole below the casing (fig. 25). Air-hammer bits (fig. 26) are used where hard rocks are at or near the surface and sufficient weight cannot be applied to a rotary bit to achieve penetration. Special fixed-head designs include antiwhirl, core, eccentric, side-tracking, special purpose (fig. 27), and steerable bits for directional drilling (fig. 28).

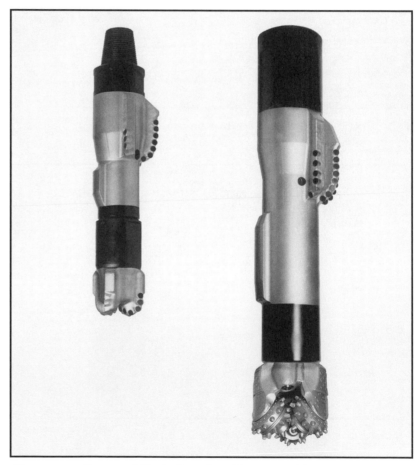

Figure 25. Ream-while-drilling bits (Courtesy of Hughes Christensen)

Figure 26. *Air hammer bits (Courtesy of Smith Bits)*

For a detailed treatment of rotary drilling bits, refer to Unit I, Lesson 2, *The Bit*, which is published by Petroleum Extension Service.

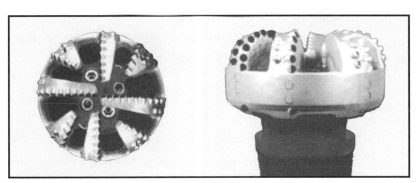

Figure 28. *Steerable directional drilling bit (Courtesy of Reed-Hycalog)*

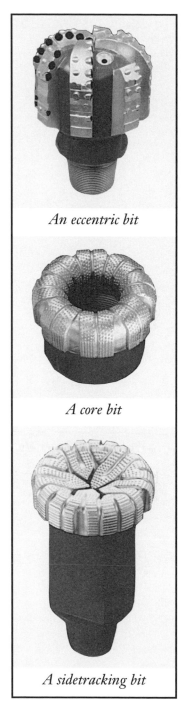

An eccentric bit

A core bit

A sidetracking bit

Figure 27. *Special purpose bits*

25

Bit Classification

Considerable confusion exists over the different ways bit manufacturers identify their products. This led the IADC to devise a standard classification system. With this system, manufacturers can indicate the characteristics of their bits, including the primary formations for which the bit is designed (table 3). By comparing charts from different suppliers, a contractor can evaluate the performance of similarly rated bits. Separate forms are available for roller cone, PDC, diamond, and TSP bits.

A four character code is used for classifying each bit. The first number is the bit series; numbers 1, 2, and 3 are for milled-tooth bits suitable for soft, medium, and hard formations, respectively. Series numbers 4 through 8 are for insert bits, also classified according to formation hardness. Thus, in the IADC system, a series 1 milled-tooth bit is classified for the same formations as a series 4 insert bit.

The second number designates the type of cutting structure and divides each series into four degrees of hardness, 1 for the softest formation in the series and 4 for the hardest.

The third character describes the bearings and gauge protection of the bit with the numbers 1 to 7. Codes 1 to 5 refer to roller bearings and 6 and 7 are for journal bearings. The chart divides bearing types into sealed, nonsealed, and air-cooled. Codes 3, 5, and 7 indicate gauge protection features.

The last character in the code indicates the most important of 16 special features a bit may have. This fourth code is optional but is useful.

In 1990 the IADC adopted a classification system for natural diamond, PDC, and TSP bits (tables 4 and 5). The codes for these bits also have four characters. The four characters indicate—

1. Body material—steel or matrix.
2. Cutter density—indicated by a number from 1 to 4 for PDC bits and 5 to 8 for TSP and natural diamond bits.
3. Cutter size or type—a number from 1 to 4 indicates the size of the PDC cutter (table 4) or type of bit, if natural diamond or TSP (table 5).

Table 3
IADC Roller Cone Bit Classification Chart

Manufacturer: _____   Date: _____

BEARING TYPE AND GAUGE SURFACE

	SERIES	TYPES	Standard Roller Bearing 1	Roller Bearing Air Cooled 2	Roller Bearing Gauge Protected 3	Sealed Roller Bearing 4	Sealed Roller Brg. Gauge Protected 5	Sealed Friction Bearing 6	Sealed Friction Brg. Gauge Protected 7
FORMATIONS									
Soft Formations with Low Compressive Strength and High Drillability	1	1 2 3 4							
Medium to Medium Hard Formations with High Compressive Strength	2	1 2 3 4							
Hard Semiabrasive and Abrasive Formations	3	1 2 3 4							
Soft Formations with Low Compressive Strength and High Drillability	4	1 2 3 4							
Soft to Medium Formations with Low Compressive Strength	5	1 2 3 4							
Medium-Hard Formations with High Compressive Strength	6	1 2 3 4							
Hard Semiabrasive and Abrasive Formations	7	1 2 3 4							
Extremely Hard and Abrasive Formations	8	1 2 3 4							

STEEL-TOOTH BITS (Series 1–3)
INSERT BITS (Series 4–8)

FEATURES AVAILABLE

A – Air Application
B – Special Bearing Seal
C – Center Jet
D – Deviation Control
E – Extended, Full-Length Jets
G – Additional Gauge and Body Protection
H – Horizontal or Steering Application
J – Jet Deflection
L – Lug Pads
M – Motor Application
S – Standard Steel Tooth Model
T – Two-Cone Bit
W – Enhanced Cutting Structure
X – Predominantly Chisel-Tooth Insert
Y – Conical Tooth Insert
Z – Other Shape Insert

1) Several features may be available on any particular bit. The fourth character should describe the predominant feature.
2) All bit types are classified by relative hardness only and will drill effectively in other formations.
3) Please check with the specific bit supplier for additional information.

Courtesy of IADC

Table 4
IADC Classification Chart for PDC Bits

(M) = Matrix Body   (S) = Steel Body

Cutters Density	Size	1 Fishtail EC	DBS	HYC	STC	SEC	2 Short EC	DBS	HYC	STC	SEC	3 Medium EC	DBS	HYC	STC	SEC	4 Long EC	DBS	HYC	STC	SEC
1	1 (>24)	R522(M) R573(M) R523(M)				B943(M)					B17-4(M)										
1	2 (14-24)		PD12(S)	DS40(S) DS33(S)	S95(S)	B933(M)															
1	3 (<14)	R423(M) AR423(M) Z426(M)	PD10(S) PD11(S)		S98(S) S93(M)	B923(M)															
2	1 (>24)	R525(M)							DS30(S)			R516(M)				B254(M)					
2	2 (14-24)	R526(M)	TD19L(M)			B925(M)				S25(S)				DS34(S)							
2	3 (<14)	R426(M) Z426(M)	TD2A1(M)	DS39(M)	S93(M)	B935(M)	R482(M)	PD1(S)	DS46(S)	S10(S)	HZ232(M) B2S(M)		LX201(M) LX101(M)	DS26(S) DS31(S)	S45(S)						
3	1 (>24)											R535(M)									
3	2 (14-24)		TD19M(M)									R535S(M)	PD4(S)								
3	3 (<14)		TD5A1(M)			B927(M)	AR435(M)	TD268(M) TD260(M)	DS23(S) DS49(M)		MX42(M)	R435(M)	PD2(S)		S85(S) S43(S)	B272(M)					
4	1 (>24)							PD5(S)					PD4HS(S)								
4	2 (14-24)		TD19H(M)									Z528(M)									
4	3 (<14)							TD290(M)			HZ352(M) B352(M)	R437(M) Z437(M)	LX401(M) LX301(M)	D247(M)	S35(M)	S292(M)	R419(M) R428(M) Z428(M)	LX271(M) TD115(M) LX291(M)	DS18(M) DS19(M) DS29(M)		B102(M) B362(M)

JC001

Courtesy of IADC

Table 5
IADC Classification Chart for TSP and Natural Diamond Bits

Size	Element	1 Flat					2 Short					3 Medium					4 Long				
	Cutters	EC	DBS	HYC	STC	SEC	EC	DBS	HYC	STC	SEC	EC	DBS	HYC	STC	SEC	EC	DBS	HYC	STC	SEC
6 <3 SPC	1 NAT											D262 D311	TB16	901 932		N37	D18			N42	
	2 TSP						S725					S225	TT16	211 241							
	3 COMB												TBT16	211ND 241ND							
7 3-7 SPC	1 NAT	D411	TB26	828		N4S	D41	TB521				D262 D331 D311	TB601	901 730 753 744		N39 N50	T51 T54	TB593 TB703	901DT		
	2 TSP	SST		828TSP				TT521	263		P443	S248 S226	TT601	243 223		P341 P343		TT593			
	3 COMB							TBT521	263ND				TBT601	243ND 223ND				TBT593 TBT703			
8 >7 SPC	1 NAT						D24		525 585		N60										
	2 TSP																				
	3 COMB																				
	4 IMP	S279						TB521													

JC002

Courtesy of IADC

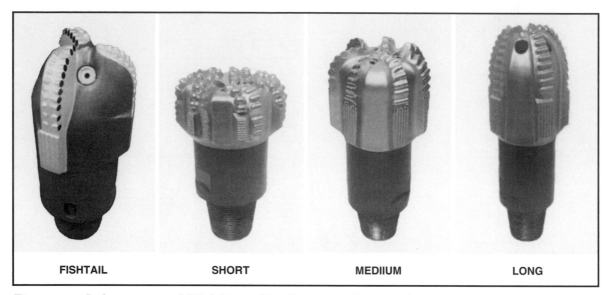

FISHTAIL   SHORT   MEDIIUM   LONG

Figure 29. Industry accepted PDC bit profiles (Courtesy of Society of Petroleum Engineers)

4. Body style and profile—a number from 1 to 4 gives a general idea of the bit profile: fishtail or flat, short, medium, and long profile (fig. 29).

Bit manufacturers provide charts that show the IADC code and list the bit models from various makers for that code (table 6). The IADC classification system gives only approximate information about a bit but provides a starting point to compare bits from different manufacturers. For example, table 6 shows a comparison for IADC code 136 (soft formation) tooth bits. It shows that a Reed HP 13, a Hughes ATJ3, a Smith FDG, and a Security S44F are equivalent.

Table 6
Tooth Bit Comparison Chart

IADC Code	Reed-Hycalog	Hughes Christensen	Smith	Security
111	Y11	R1	DSJ	S3SJ, S3SJD, 2S3JD
114		ATX1, GTX1, X3A	SDS	S33S, MS33S, SS33S
115	EMS11G, EMS11DH, MS11G, MS11DH	ATXG1, MAXG1, MAXGT1, GTXG1	MSDSH, MSDSHOD, MSDSSH	S33SG, SS33SG
116	HP11, HP11+, EHT11	ATJ1, ATJS, ATM1, ATM1S, ATM1H, GT1, ATJ1S	FDS, FDS+, FDSS+	S33SF, S33SFX, PSF
117	MHT11G, MHT11DH	ATMG1, ATMG1S ATJG1H, GTG1, GT1H, ATMGT1	MFDSH, MFDSSH, MFDSHOD	S33SGF
121	Y12	R2 DTJ	S3J, S3TJ	
124				S33
125				S33G, SS33G
126	EHP12, HP12, EHT12	ATJ2, J2, J2T	FDT	S33F
127		JG2	S33TGF	
131	Y13	R3 DGJ	S4J, S4TJ, S4T	
134	ATX3, GTX3	S44		
135	EMS13G, EMS13DH, MS13G, ETS 13G	ATXG3, MAXG3, MAXGT3, GTXG3	SDGH, MSDGH, MSDGHOD	S44G, SS44G
136	HP13	ATJ3	FDG	S44F
137	HP13G, MHT13G, MHT13DH	JG3, ATMG3, ATMGT3 MFDSHOD	FDGH, MFDGH,	S44GF
211		R4 V2J	M4NJ	
214				M44N
215			SVH, MSVH	M44NG, MM44NG
216	HP21	J4, ATJ4	FV	M44NF
217	HP21G	JG4	FVH	M44NGF
221		DR5		
415	EMS41H, EMS41HDH	ATX05, MAX05, GTX03, MAX03, MAX00, MAX00, MAXGTOD, MAXGT03	M02SOD, M015 M025, M025OD, 02M	SS80

Table 6, cont.
Tooth Bit Comparison Chart

IADC Code	Reed-Hycalog	Hughes Christensen	Smith	Security
417	EHP41, EHP41H, EHP41A, EHP41ADH, HP41A	ATJ00, ATM00, ATJ05 ATM05, GT03, ATMGT03, GT00, ATMGT00	F05	S80F
425		ATX05C, MAX05C	M05S, 05M, 05MD	SS81
427		ATJ05C, ATM05C, GT05, GT03C	F07, F05, MF05, 05MF, 05MFD	S91F
435	EMS43A, EMS43ADH, MS43A-M, MS43AD-M	ATX11H, ATX11S, MAX11H, ATX11, GTX09, MAXGT09	M1S, M15OD, 10M, 10MD, 12M, 12MD, 12MY	S82, SS82
437	EHP43, EHP43A, EHP43H, EHP43ADH, EHP43HDH, HP43, HP43A, HP43A-M HP43-M	ATJ11H, ATM11, ATM11H, ATMGT09, ATJ11, ATJ11S, ATM11HG, GT09	F1, F1S, MF1, F10D, MF10D, 10MF, 10MFD, 12MFD, 12MF, 12MFY	S82F, S82CF, SS82F
445	EMS44H, EMS44HDH, EMS44A, EMS44ADH	ATX11C, MAX09	15JS, M15SD, M15S, M15SOD, 15M, 15MD	
447	HP44-M, EHP44H, EHP44HDH, HP44	ATM11C, ATM11CG, ATM18, ATJ11C, ATJ18, GT18C, GT09C, ATMGT18	F15, F15S, F17, MF15, MF15D, MA15, MF15, 15MF, MF15OD, MF15D, 15MFD	S83F, SS83F
515	EMS51A, EMS51H, EMS51ADH, MS51A, MS51A-M, MS51ADHM, ETS51, ETS51H	ATX22, MAX22, GTX18, MAXGT18, MAX22G	2JS, M2S, M2SD, A1JSL, MA1SL	S84, SS84
517	EHP51, EHP51A, EHP51H, EHP51ADH, EHP51HDH, EHP51X EMS51A, HP51, HP51A, HP51A-M, HP51H, HP51H-M, HP51X-M, · HP51X	ATJ22, ATJ22G, ATM22, ATM22G, ATJ22S, GT20, GT18, ATMGT18, ATMGT20	F2, F2H, F25, A1, F15H, F17, F25A, MF2D, MF2, F2D	S84F, S84CF DS84F, SS84F
525			M27S, M27SD	SS85
527	HP52, HP52A, HP52X, HP52-M	ATJ22C, ATM22C, ATM28, ATJ28, GT28, GT18C, GT20C, GT28C, ATJ28C	F27, MF27D, F27i, MF27	S85F, S85CF
535	EMS53A, MS53, MS53DH	ATX33, ATX33A	3JS, M3S, M3SOD	S86, SS86
537	EHP53, EHP53A, EHP53DH, EHP53ADH, HP53, HP53A, HP53A-M, HP53ADH,	ATJ33, ATJ33S, ATJ33A, ATM33, ATJ35, ATJ33D, ATJ33G, ATJ33H	F3, F3H, MF3H, MF3D, MF3OD, MF3, F3D	S86F, S86CF, SS86F

Courtesy of Reed-Hycalog

The decision to change bits can greatly influence the cost of drilling. A bit pulled too soon means extra trip time (about 1 hour/1,000 feet or 300 metres at moderate depths) and inefficient use of the bit; pulled too late means hours of less-than-optimum penetration and the danger of cone loss that could require fishing time.

All bits wear out. The primary clues to bit condition include rotating hours, slowing of penetration rate, torque variations, jerky rotary action, and pump pressure changes. Decreasing penetration rate is the most common reason to change bits. Other common reasons to pull a bit include an increase in torque, a change of formation, junk in the hole, and reaching a testing, casing, or logging depth.

Operations such as logging, testing, or coring often influence a decision to change bits because the pipe is already out of the hole. Thus, crew members may sometimes run a new bit while there is still significant wear remaining on the old bit (the old bit is still "green"). For instance, if a DST is run only 150 feet (50 metres) above an expected change to a harder formation, a hard formation bit may be run early in order to save a trip a few hours later. Similarly, mechanical problems, on the rig or downhole, often dictate a bit change at an unscheduled moment.

An operator's well plan may include a bit program or the company may request a bit manufacturer to furnish a recommendation. Bit companies usually have extensive information on nearby wells and can provide computer-generated programs that lay out the planned bit-change depths, bit selection, recommended weight and rpm, and cost data (see fig. 4). If followed, the recommended program can predetermine the bit-change schedule.

Once pulled, an examination of the bit by the driller, rig manager, or bit company representative can reveal useful information about how the bit performed. Proper grading of dull bits is important because it can indicate what changes, if any, are needed in the company's drilling program. Good dull-bit evaluations can help guide a bit maker's design of future products by revealing problems and ways to correct them.

Dull Bit Evaluation
When to Pull the Bit

IADC Dull Bit Grading System

The IADC has devised a standardized system for grading dull bits (table 7). The system is used for both roller and fixed-cutter (diamond) bits. Both types of bit are graded on eight categories of wear. Since fixed-cutter bits have no bearings, the column for bearing wear (B) is marked with an X. The list of cutter wear codes includes some that apply only to roller cone bits and some that apply only to fixed-head cutter wear.

Dull bits are graded on the basis of cutter wear, bearing wear (not for diamond and PDC bits), and gauge wear. The first four columns refer to the condition of the cutters and their location on the cone or bit body (for both roller cone and fixed-cutter bits). Cutter wear can be measured directly, compared to the new condition, then given an accurate grade of 0 to 8 (fig. 30). Select the appropriate code(s) for the dull character of the bit and record in column 3, and also under Remarks, column 7.

Bearings and seals in column B (5), whether sealed or nonsealed, are more difficult to grade and are estimates, at best. Only experienced drillers can make a good estimate of bearing wear. Note that sealed and nonsealed bearings use different codes—numbers 0–8 for nonsealed and E, F, and N for sealed bearings.

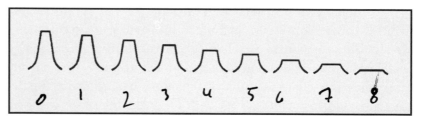

Figure 30. Cutter wear measurement (Courtesy of Society of Petroleum Engineers)

Table 7
IADC Dull Grading Chart

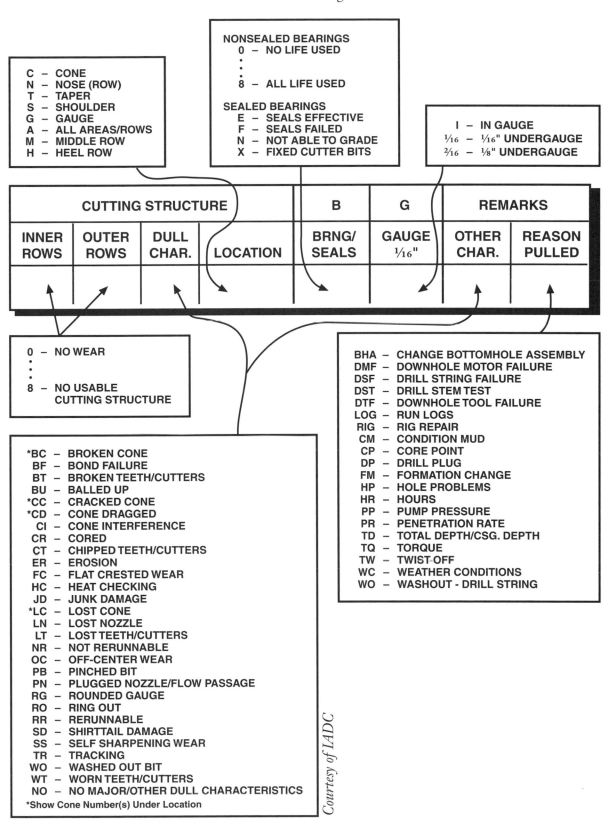

C — CONE
N — NOSE (ROW)
T — TAPER
S — SHOULDER
G — GAUGE
A — ALL AREAS/ROWS
M — MIDDLE ROW
H — HEEL ROW

NONSEALED BEARINGS
0 — NO LIFE USED
.
.
.
8 — ALL LIFE USED

SEALED BEARINGS
E — SEALS EFFECTIVE
F — SEALS FAILED
N — NOT ABLE TO GRADE
X — FIXED CUTTER BITS

I — IN GAUGE
1/16 — 1/16" UNDERGAUGE
2/16 — 1/8" UNDERGAUGE

CUTTING STRUCTURE				B	G	REMARKS	
INNER ROWS	OUTER ROWS	DULL CHAR.	LOCATION	BRNG/ SEALS	GAUGE 1/16"	OTHER CHAR.	REASON PULLED

0 — NO WEAR
.
.
.
8 — NO USABLE
    CUTTING STRUCTURE

*BC — BROKEN CONE
 BF — BOND FAILURE
 BT — BROKEN TEETH/CUTTERS
 BU — BALLED UP
*CC — CRACKED CONE
*CD — CONE DRAGGED
 CI — CONE INTERFERENCE
 CR — CORED
 CT — CHIPPED TEETH/CUTTERS
 ER — EROSION
 FC — FLAT CRESTED WEAR
 HC — HEAT CHECKING
 JD — JUNK DAMAGE
*LC — LOST CONE
 LN — LOST NOZZLE
 LT — LOST TEETH/CUTTERS
 NR — NOT RERUNNABLE
 OC — OFF-CENTER WEAR
 PB — PINCHED BIT
 PN — PLUGGED NOZZLE/FLOW PASSAGE
 RG — ROUNDED GAUGE
 RO — RING OUT
 RR — RERUNNABLE
 SD — SHIRTTAIL DAMAGE
 SS — SELF SHARPENING WEAR
 TR — TRACKING
 WO — WASHED OUT BIT
 WT — WORN TEETH/CUTTERS
 NO — NO MAJOR/OTHER DULL CHARACTERISTICS
*Show Cone Number(s) Under Location

BHA — CHANGE BOTTOMHOLE ASSEMBLY
DMF — DOWNHOLE MOTOR FAILURE
DSF — DRILL STRING FAILURE
DST — DRILL STEM TEST
DTF — DOWNHOLE TOOL FAILURE
LOG — RUN LOGS
RIG — RIG REPAIR
CM — CONDITION MUD
CP — CORE POINT
DP — DRILL PLUG
FM — FORMATION CHANGE
HP — HOLE PROBLEMS
HR — HOURS
PP — PUMP PRESSURE
PR — PENETRATION RATE
TD — TOTAL DEPTH/CSG. DEPTH
TQ — TORQUE
TW — TWIST-OFF
WC — WEATHER CONDITIONS
WO — WASHOUT - DRILL STRING

Courtesy of IADC

Column G (6) is used to record whether the bit can still drill a full-gauge hole. This is an important measurement because an undergauge hole requires expensive reaming time. For tricone bits, use a gauge ring and the two-thirds rule to record the amount of undergauge to the closest 1/16-inch (in.) or millimetre (mm) (fig. 31). Measure the distance from the ring to the third cone and multiply it by 2/3 (0.6666) then record the result to the nearest 1/16 in. (mm) in column G. If the hole is seriously out of gauge, special care while going back to bottom and low weight on the bit is recommended. Reamers may need to be selectively placed in the drill string.

The last column is used to record the reason the bit was pulled.

The parts that usually fail in roller cone bits are the cone, the teeth, and the bearings (fig. 32).

Cone Wear

Cone skidding, or dragging, occurs when a cone stops turning as the bit rotates (fig. 33). The usual causes of cone skidding are locked cones that result in bearing failure. Excessive weight or rotary speed is the usual cause of bearing failure. Junk in the hole, a balled up, or a pinched bit can also cause cone skidding.

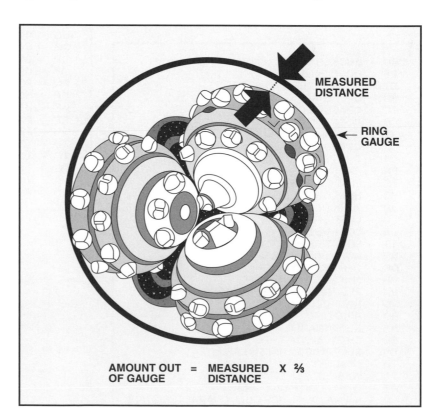

MEASURED
DISTANCE

RING
GAUGE

AMOUNT OUT   =   MEASURED   X 2/3
OF GAUGE   DISTANCE

Figure 31. Gauge ring measurement using the two-thirds rule

Figure 32. An evenly worn dull bit, which is in gauge. (Courtesy of Hughes Christensen)

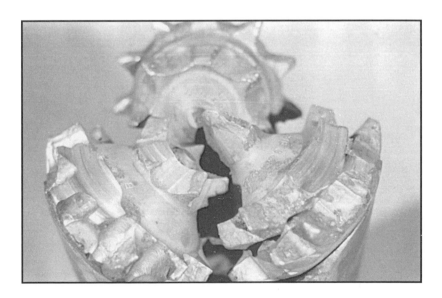

Figure 33. Cone dragging or skidding caused by a locked cone

Cone interference is a condition in which the cones or cutters run into each other. Interference occurs when the cones are pinched in by too much weight while reaming or by bearing wear. The result of cone interference is broken teeth or cracked cones (fig. 34) or a cut off nose of a cone.

Cracked cones may stay attached to the bit; if, however, they break into two or more pieces and fall off they must be retrieved or milled before drilling ahead. Dropping the drill string or hitting the bit on a ledge during a trip can break cones. Cone interference can also cause cones to break.

Cone erosion is the wearing away of the cone body (fig. 35). It is more common in air or gas drilling where the circulating fluid is charged with abrasive cuttings and is moving at a very high velocity, producing a sand-blasting effect. Severe cone erosion can cause cracked or broken cones and loss of inserts.

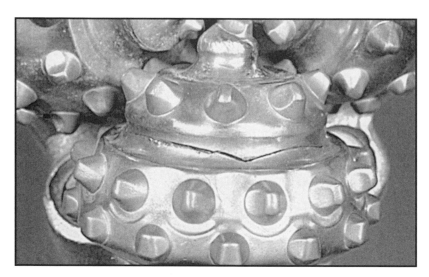

Figure 34. Cracked cone

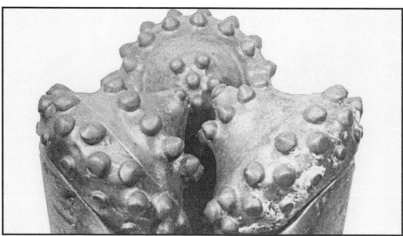

Figure 35. Cone erosion

Off-center wear often occurs when the rate of penetration is too slow because of slow rpm or insufficient weight on the bit. When penetration is too slow, the cone offset can cause the bit to rotate off-center (whirling). This is a condition where two cones drill the bottom of the hole and one drills the side. The cone drilling the side will show excessive gauge area wear (fig. 36). Off-center bit wear is also common when downhole motors are used with a directional bottomhole assembly (bent sub). Excessive mud weight and a bit too hard for the formation can also cause off-center wear.

Center coring is a condition in which the inside rows of cutters wear, are lost, or the nose of a cone is worn away. Junk in the hole or cone erosion is the usual cause of center coring (fig. 37). Improper break-in can cause center coring so it is important that the first few feet (metres) be drilled light and slow.

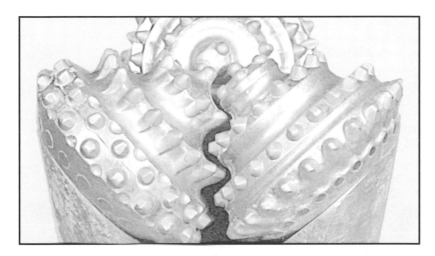

Figure 36. Off-center wear

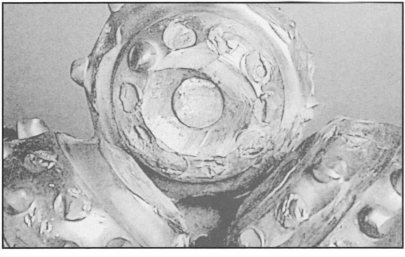

Figure 37. Center coring

Tooth Wear

Insert Bits. Broken inserts slow penetration and are a common problem; however, it is normal in some formations (fig. 38). Often, the wrong bit was selected if broken inserts occur early and appear excessive. Longer inserts used in soft formations cannot withstand the shock encountered in harder rock. Other causes of broken inserts are junk in the hole, high rotary speeds, excessive weight, improper break-in, slamming into a ledge or the bottom of the hole, and cone interference.

Excessive wear or broken inserts on the outside rows of hard-formation bits indicates that the rotary speed was too high.

Insert loss is usually the result of a cone problem—erosion, cracking, interference, or corrosion. Once lost, inserts on the bottom of the hole nearly always cause further damage to the bit because they are so hard. Being dense (heavy), they are difficult to circulate out of the hole and must usually be recovered in a junk sub.

Repeated dragging (heating) and cooling of the inserts causes heat checking or cracking (fig. 39). A lower rpm may correct the problem.

Steel-tooth bits. Steel teeth will often be chipped. Cone interference and rough formation conditions cause chipped teeth. On the other hand, tooth breakage on a milled-tooth bit is not normal and indicates the wrong bit was used (long tooth instead of short, for instance) or improper weight on bit (WOB) or rpm (fig. 40).

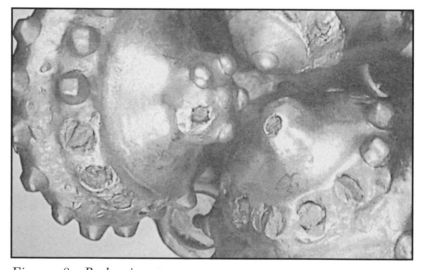

Figure 38. Broken inserts

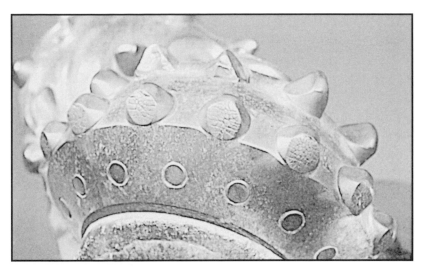

Figure 39.   Heat checking (small cracks) on insert cutters

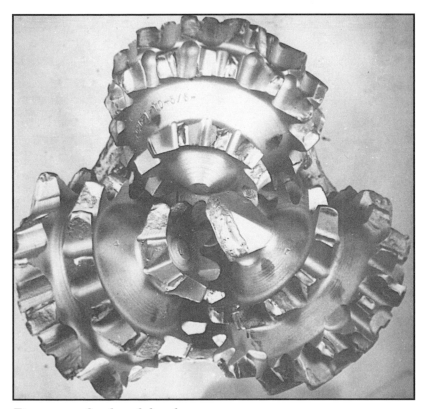

Figure 40.   Steel tooth breakage

Self-sharpening and flat-crested wear are two types of normal tooth abrasion. Self-sharpening wear is desirable and can be designed into the bit by placing hard metal on one side of each tooth and softer material on the other side. The rock abrades the softer metal on each tooth thus keeping them sharp (fig. 41). Self-sharpening is seen less now that full-tooth hardfacing is in common use. In flat-crested wear, the teeth wear evenly and flat.

Bradding usually affects the inner rows of teeth. It occurs when excessive weight cracks the hardfacing on the teeth and exposes the softer metal underneath.

Tracking is a condition usually confined to plastic shale and seen more often in smaller bits or with downhole motors. The pattern on the bottom of the hole matches the tooth pattern and they mesh together like gears, greatly slowing penetration. A longer tooth bit may solve the problem.

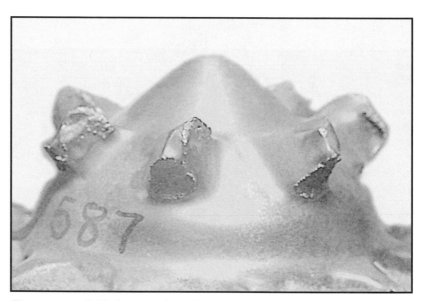

Figure 41. Self-sharpened tooth wear

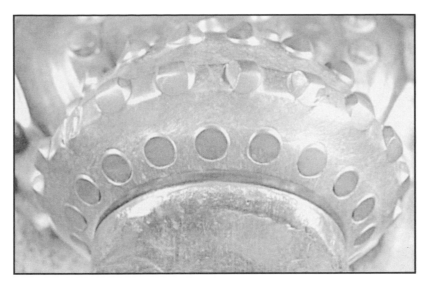

Figure 42. Bit gauge wear or rounding

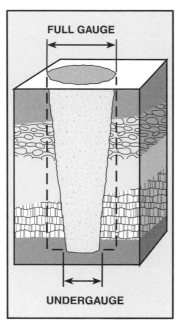

Figure 43. Undergauge hole caused by bit gauge wear or rounding

Excessive gauge wear can result from an unstable drill string (too few or too light collars) or high rpm. Gauge wear or rounding (fig. 42) in any bit is a serious problem if the bit no longer drills a full-size hole (fig. 43). The next bit will have to ream to bottom, possibly causing gauge wear to that bit. Using a bit with greater gauge protection can help correct the problem.

Bearing Wear

Bearing wear and failure often make it necessary to replace the bit. Sealed bearings and improved metallurgy extend bit life but have not eliminated bearing wear. Both the journal and roller bearings can fail if the WOB or the rotary speed is too high, an abrasive drilling fluid is used, the bit is used for reaming an undergauge hole, or is run too long.

The *outer bearings* of a bit are the large bearings under the gauge cutters. They may fail for various reasons—too many hours or excessive weight are common causes. Skid marks on the cones or locked cones are usual signs of outer bearing failure.

The *inner bearings* are the small inside ball bearings and the nose bearing. They fail when drilling a formation that is too hard for the bit. Flat-crested tooth wear is a sign of inner bearing wear.

43

To summarize—

- Geologic prognosis aids in bit selection for a particular well.
- MWD provides information while hole is being drilled.
- Type of formation is the most important factor in bit design.
- Bit types include:
  - Roller cone, or rock, bits
    - Steel tooth bits
    - Tungsten carbide insert bits
  - Fixed-cutter, or fixed-head, bits
    - Natural diamond bits
    - PDC bits
- Steel-tooth and tungsten carbide bits are available for drilling soft, medium soft, medium, medium hard, hard, and very hard formations.
- Soft formation roller cone bits have more offset than hard formation bits.
- Soft formation roller cone bits have fewer and longer cutters than hard formation bits.
- Natural diamond bits are fixed-head bits in which several industrial-grade diamonds are embedded.
- A diamond bit's shape and the size, number, and placement of the diamonds determine a diamond bit's ability to drill a particular formation.
- PDC bits have cutters made of synthetic diamond crystals bonded to a tungsten carbide insert. The insert is brazed into pockets in the bit head or onto blades.
- Thermally stable polycrystalline (TSP) diamond bits are more stable than PDC bits at high temperatures.
- Hybrid bits combine the best qualities of natural diamond, PDC, TSP, or tungsten carbide cutters into one bit.

Drilling Performance Records

Accurate drilling performance records are of great value. They directly influence the selection of bits and help determine proper operating procedures that impact total drilling costs.

Bit records show many critical facts about the operation (fig. 44). The most important statistics describe each bit run, including bit type, drilling hours, footage drilled, nozzle size, reason pulled, and dull condition. Other information shown on the bit record includes the rotary weight and speed, circulation system, and deviation, all of which affect cost calculations.

The *daily drilling report* is another important performance record (fig. 45). The report is a 24-hour record of the drilling operation and provides a complete and accurate account of the drilling progress by each tour. The driller signs and is responsible for the report on his tour. The daily drilling report helps achieve consistency in the drilling process because each driller can refer to the progress made by the last tour. Continuity of operations is assured because the report sets out the current conditions or problems assumed by the new crew, including bit performance, mud program, drilling assembly, and time required for various rig operations or mechanical problems. The operator will usually extract information from the rig's daily report and condense it for office use (fig. 46).

The increasing use of computers in integrated drilling systems allows instant communication between the drill site and multiple remote stations (fig. 47). Measurement-while-drilling systems (MWD) may steer a downhole motor or relay data from downhole instruments that measure formation properties (see fig. 4).

895- 32803

API WELL NO. ____

CONTRACTOR ____ RIG NO. ____

TOOLPUSHER ____

OIL COMPANY ____

OIL COMPANY REPRESENTATIVE ____

REED

TOOL COMPANY

☐ DAYWORK
☐ FOOTAGE TO ____
☐ TURNKEY

REED REP. ____ DIST. ____

SURVEY ____ ABS ____ BLOCK ____

SECTION ____ TOWNSHIP ____ RANGE ____

COUNTY ____ STATE ____

LEASE ____ WELL NO. ____

MUD CO. ____ MUD TYPE/DEPTH ____

RIG MAKE AND MODEL ____

NO. 1 PUMP MAKE AND MODEL ____

NO. 2 PUMP MAKE AND MODEL ____

SPUD DATE ____

UNDER SURFACE ____

INTERMEDIATE ____

T. D. DATE ____

DRILL PIPE : SIZE/TYPE ____ HEAVY WEIGHT : SIZE/TYPE ____ LENGTH ____

DRILL COLLARS: NO. ____ O.D. ____ I.D. ____ LENGTH ____

DRILL COLLARS: NO. ____ O.D. ____ I.D. ____ LENGTH ____

BIT NO.	BIT SIZE	BIT MFGR	BIT TYPE	BIT SERIAL NO.	JET SIZE OR TFA	DEPTH OUT	FTGE	HOURS RUN	ACC HOURS	FT/HR	WOB	RPM	VERT DEV	PUMP PRESS	PUMPS			MUD			DULL CODE							REASON PULLED	DATE (REMARKS)
															NO.	LIN	SPM	WT	VIS	WL	CUTTING STRUCTURE T_I	T_O	MDC	Loc	B	GAUGE	ODC		

DULL CODE DEFINITIONS

TOOTH CONDITION

T_I = INNER ROWS
T_O = OUTER ROWS

0 = NO WEAR
1 = LESS THAN 1/8 TOOTH WEAR
8 NO USABLE TOOTH HEIGHT

MDC = MAJOR DULL CHARACTERISTICS
ODC = OTHER DULL CHARACTERISTICS

*BC – BROKEN CONE
BT – BROKEN TEETH/CUTTERS
BU BALLED UP
*CC – CRACKED CONE
*CD – CONE DRAGGED
CI – CONE INTERFERENCE
CR – CORED
CT – CHIPPED TEETH
ER – EROSION
FC – FLAT CRESTED WEAR
HC HEAT CHECKING
JD – JUNK DAMAGE
*LC – LOST CONE
LN – LOST NOZZLE
LT – LOST TEETH/CUTTERS
OC – OFF CENTER WEAR
PB – PINCHED BIT
PN – PLUGGED NOZZLE
RG – ROUNDED GAUGE
RO – RING OUT
SD – SHIRTTAIL DAMAGE
SS – SELF SHARPENING WEAR
TR – TRACKING
WO – WASH OUT ON BIT
WT – WORN TEETH/CUTTERS
NO – NO OTHER MAJOR/OTHER DULL CHARACTERISTIC

Loc = LOCATION

OF CUTTING STRUCTURE
DULL CHARACTERISTIC
ROLLER CONE
N–NOSE ROWS — CONE # OR #'s
M–MIDDLE ROWS — 1
G–GAUGE ROWS — 2
A–ALL ROWS — 3
FIXED CUTTER (DIAMOND)
C–CONE — S–SHOULDER
N–NOSE — G–GAUGE
T–TAPER — A–ALL AREAS

B = BEARING DULL CODE

NONSEALED BEARINGS
0–NO LIFE USED
•
•
8–ALL LIFE USED

SEALED BEARINGS
E–SEALS EFFECTIVE
F–SEALS FAILED

X–FIXED CUTTER BITS

GAUGE WEAR

I–IN GAUGE
1/16–1/16" UNDERGAUGE
2/16–1/8" UNDERGAUGE
•
•
TO NEAREST 1/16TH OF AN INCH

REASON PULLED

BHA - CHANGE BOTTOM HOLE ASSEMBLY
DMF– DOWN HOLE MOTOR FAILURE
DSF– DRILL STRING FAILURE
DST– DRILL STEM TEST
DTF– DOWN HOLE TOOL FAILURE
LOG– RUN LOGS
CM– CONDITION MUD
CP– CORE POINT
DP– DRILL PLUG
FM– FORMATION CHANGE
HP– HOLE PROBLEMS
HR– HOURS ON BIT
PP– PUMP PRESSURE
PR– PENETRATION RATE
RIG– RIG REPAIRS
TD– TOTAL DEPTH/CASING DEPTH
TQ– TORQUE
TW TWIST OFF
WC– WEATHER CONDITIONS
WO– WASHOUT–DRILL STRING

Figure 44. Bit record form (Courtesy of Reed-Hycalog)

TIME DISTRIBUTION – HOURS

CODE NO. – OPERATION	MORN.	DAY	EVE.
1. RIG UP AND TEAR DOWN			
2. DRILL ACTUAL			
3. REAMING			
4. CORING			
5. CONDITION MUD & CIRCULATE			
6. TRIPS			
7. LUBRICATE RIG			
8. REPAIR RIG			
9. CUT OFF DRILLING LINE			
10. DEVIATION SURVEY			
11. WIRE LINE LOGS			
12. RUN CASING & CEMENT			
13. WAIT ON CEMENT			
14. NIPPLE UP B.O.P.			
15. TEST B.O.P.			
16. DRILL STEM TEST			
17. PLUG BACK			
18. SQUEEZE CEMENT			
19. FISHING			
20. DIR. WORK			
21.			
22.			
23.			

COMPLETION

	MORN.	DAY	EVE.
A. PERFORATING			
B. TUBING TRIPS			
C. TREATING			
D. SWABBING			
E. TESTING			
F.			
G.			
H.			
TOTALS			

DAYWORK TIME SUMMARY (OFFICE USE ONLY)

HOURS W/CONTR. D.P.		
HOURS W/OPR. D.P.		
HOURS WITHOUT D.P.		
HOURS STANDBY		
TOTAL DAYWORK		
NO. OF DAYS FROM SPUD		
CUMULATIVE ROTATING HOURS		
DAILY MUD COST		
TOTAL MUD COST		

DRILLING ASSEMBLY (At end of tour)

NO.	ITEM	LENGTH
	BIT	
	OD	
	OD	
	OD	
	OD	
	OD	
	STANDS ___ D.P.	
	SINGLES ___ D.P.	
	KELLY DOWN	
	TOTAL	
WT. OF STRING		

BIT RECORD

BIT NO.		
SIZE		
IADC CODE		
MANUFACTURER		
TYPE		
SERIAL NO.		
JETS		
TFA		
DEPTH OUT		
DEPTH IN		
TOTAL DRILLED		
TOTAL HOURS		

CUTTING STRUCTURE

INNER	OUTER	DULL CHAR.	LOCATION

BEARINGS/SEALS	GAGE	OTHER DULL CHAR.	REASON PULLED

MUD RECORD

	TIME			
WEIGHT				
PRESSURE GRADIENT				
FUNNEL VISCOSITY				
PV/YP				
GEL STRENGTH				
FLUID LOSS				
pH				
SOLIDS				

MUD & CHEMICALS ADDED

TYPE	AMOUNT	TYPE	AMOUNT

(MORNING TOUR / DAY TOUR / EVENING TOUR — the three identical DRILLING ASSEMBLY, BIT RECORD, and MUD RECORD blocks repeat for each tour)

© 1995 International Association of Drilling Contractors.

IADC - API OFFICIAL DAILY DRILLING REPORT FORM

APPROVED   APPROVED

PRINTED IN U.S.A.

Figure 45. Rig site daily drilling report form (Courtesy of IADC)

XYZ Corporation
0000 1st, Suite 100
Tulsa, OK 74100
918 555 1212 ofc
918 555 0000 fax

Foster 19-1
C NW/4
Section 19-13N-04W
Oklahoma County, Oklahoma
KB Elevation, 1533'

WELL REPORT

01/14/99   Settled the surface damages.

01/15/99 thru 01/19/99   Building location.

01/20/99   Building location. Set the conductor and drilled the rat and mouse hole.

01/21/99   Started moving in the rig and rigging up.

01/22/99   Rigging up.

01/23/99   Finished rigging up. Spud at 2 p.m. on 1/22/99.
Drilling at 1523', (1523', 14 hrs), day 1, mw 9.2, vis 32, wl n/c, bit 1, 12¼", ½° at 500', ½° at 1011', $101,633/$101,633

01/24/99   Depth 1523', (0'), day 2, bit 1, 12¼". Ran 34 joints, 9⅝", 36#, J-55, ST&C, casing set at 1523.69', cemented with 446 sacks, 35/65 POZ containing 2% CC, 6% gel and ¼ pps cello-flake and 130 sacks of class C containing, 2% CC, and ¼ pps cello-flake. Plug down at 7:05 p.m., cut off casing and welded on head. $37,819/$139,452

01/25/99   Waited on cement 6 hours. Installed bop in 5½ hours. Ran bit, dc and dp and tested bop to 1500 psi in 2½ hours. Drilling at 1,981', (458', 7.00 hrs), day 3, mw 9.1, vis 31, wl n/c, bit 2, 7⅞", $13,765/$153,217

01/26/99   Drilling at 3,589', (1608', 22.00 hrs), day 4, mw 9.1, vis 32, wl n/c, pH 11.5, bit 2, 7⅞", 2¼° at 2003', $32,962/$186,179

01/27/99   Drilling at 4,857', (1273', 23.50 hrs), day 5, mw 9.1, vis 35, wl 72, pH 10.0, bit 2, 7⅞", 0° at 3682', $25,954/$212,133

01/28/99   Drilling at 6,147', (1290', 22.75 hrs), day 6, mw 9.2, vis 35, wl 42, pH 10.5, bit 2, 7⅞", ¾° at 4857', 1° at 5868', $27,953/$240,086

01/29/99   Drilling at 6,885', (738', 21.50 hrs), day 7, mw 9.3, vis 35, wl 62, pH 10.5, bit 2, 7⅞", 1¼° at 6861', $16,161/$256,247

01/30/99   Drilling at 7,034', (149', 16.50 hrs), day 8, mw 9.3, vis 34, wl 68, pH 10.0, bit 3, 7⅞", 7½ hours tripping. $5,300/$261,547

01/31/99   Drilling at 7,230', (196', 16.00 hrs), day 9, mw 9.3, vis 35, wl 56.0, pH 10.0, bit 3, 7⅞", 6½ hours tripping. $4,698/$266,245

02/01/99   Drilling at 7,525', (295', 23.75 hrs), day 10, mw 9.3, vis 35, wl 56.0, pH 10.5, bit 4, 7⅞", $7,453/$273,698

02/02/99   Drilling at 7,790', (265', 23.75 hrs), day 11, mw 9.3, vis 33, wl 57.0, pH 11.0, bit 4, 7⅞", $5,945/$279,643

02/03/99   Drilling at 7,909', (119', 14.75 hrs), day 12, mw 9.3, vis 37, wl 26.0, pH 11.0, bit 4, 7⅞", 1° at 7834', $4,422/$284,065

02/04/99   Drilling at 8,180', (271', 23.75 hrs), day 13, mw 9.3, vis 36, wl 26.0, pH 10.5, bit 5, 7⅞", $6,064/$290,129

02/06/99   Drilling at 8,528', (215', 17.25 hrs), day 15, mw 9.3, vis 38, wl 16.0, pH 10.0, bit 5, 7⅞", $4,947/$302,832

02/07/99   Drilling at 8,705', (177', 23.75 hrs), day 16, mw 9.3, vis 38, wl 14.0, pH 10.5, bit 6, 7⅞", $4,189/$307,021

02/08/99   Drilling at 8,855', (150', 14.50 hrs), day 17, mw 9.4, vis 50, wl 12.0, pH 9.5, but 6, 7⅞", $3,651/$310,672

02/09/99   Drilling at 9,039', (184', 23.75 hrs), day 18, mw 9.4, vis 41, wl 12.4, pH 10.5, bit 7, 7⅞", $5,499/$316,171

02/10/99   Drilling at 9,220', (181', 23.75 hrs), day 19, mw 9.4, vis 49, wl 12.0, pH 10.5, bit 7, 7⅞", $5,399/$321,570

02/11/99   Drilling at 9,418', (198', 22.75 hrs), day 20, mw 9.4, vis 48, wl 11.0, pH 10.6, bit 7, 7⅞", $6,218/$327,788

Questions? Call Summa Engineering Inc. at (405) 232-8338.

Figure 46. Office drilling report (Courtesy of Summa Engineering Inc.)

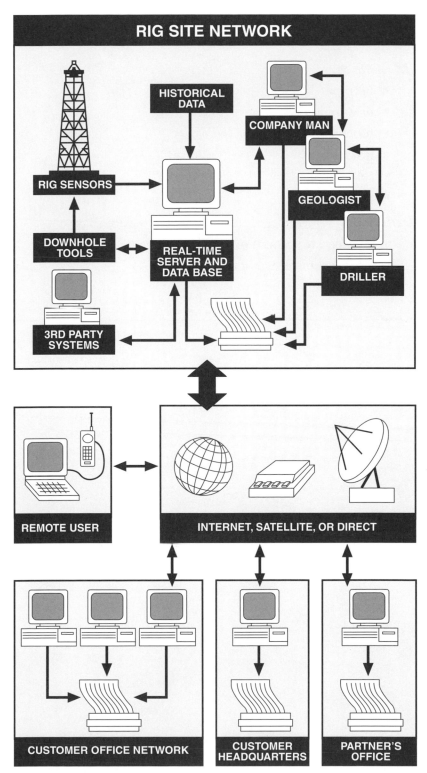

Figure 47. Computerized drilling information network

To summarize—

Bit records show

- bit type
- drilling hours
- footage (metreage) drilled
- nozzle size
- reason pulled
- dull condition

Daily drilling reports give

- an accurate account of drilling progress per tour
- a consistent record of drilling progress
- a continuity in operations

Weight on Bit and Rotary Speed

The mechanical factors of bit weight and rotary speed must be coordinated with bit selection to achieve optimal drilling rates. Generally, an increase in either weight or rpm increases the rate of penetration, provided the right bit is in place and bottomhole cleaning is attained through proper bit hydraulics. However, weight and rpm cannot be increased indiscriminately without considering other factors. The extra wear imposed on the bit bearings and cutting structures must be considered. For instance, drill string failure is more common at high rotary speed. The increase in shock loads can shatter bit teeth, especially if the formation contains both soft and hard layers. One extra trip can exceed the costs of a few hours of slower penetration. Hole deviation may also become a problem with increased weight unless the drill string is not stiff and well stabilized. Also, as the bit becomes dull, hole deviation tends to increase. When all factors are considered, simply increasing the weight or rotary speed will not always result in least-cost drilling.

With new automatic driller technology it is possible to maintain a steady weight on the bit. This technology enhances penetration rates significantly and increases both the bit life and number of reusable bits. Drilling with automatic drillers can control weight, differential pressure, or rotary amps (torque), all of which contribute to lower drilling costs per foot.

Weight and rotary speed are usually applied in inverse proportion—that is, higher speeds require lower weights and vice versa. Table 8 gives suggested weight and speed ranges for roller cone bits in formations of different hardness. The shearing action of a long-tooth soft formation bit is well-suited for high rotary speeds.

Table 8
Typical Weights and RPMs for Roller Cone Bits

IADC Code	psi of diameter (dN/mm of dia.)	rpm Range	IADC Code	psi of diameter (dN/mm of dia.)	rpm Range
116	2,000-5,000 (890-2,225)	300-80	517	2,000-5,500 (890-2,448)	120-60
117	2,000-5,000 (890-2,225)	300-80	517	2,000-6,000 (890-2,670)	120-50
137	2,000-5,000 (890-2,225)	300-80	527	2,000-6,000 (890-2,670)	120-50
417	2,000-5,000 (890-2,225)	280-70	537	3,000-5,500 (1,335-2,448)	110-50
437	2,000-5,000 (890-2,225)	280-60	537	3,000-6,000 (1,335-2,670)	110-40
437	2,000-5,500 (890-2,448)	240-60	537	3,000-5,000 (1,335-2,225)	80-40
447	2,000-5,500 (890-2,448)	240-60	547	3,000-6,000 (1,335-2,670)	100-40
517	2,000-6,000 (890-2,670)	240-50	547	3,500-7,000 (1,558-3,115)	80-40
527	2,000-6,000 (890-2,670)	240-50	617	3,000-6,000 (1,335-2,670)	80-40
537	3,000-6,000 (1,335-2,670)	220-40	617	3,000-5,500 (1,335-2,448)	80-40
547	3,000-6,000 (1,335-2,670)	200-40	627	3,500-7,000 (1,558-3,115)	80-40
116	2,000-5,000 (890-2,225)	180-80	637	3,000-6,000 (1,335-2,670)	70-35
126	2,000-5,500 (890-2,448)	150-70	637	4,000-7,000 (1,780-3,115)	70-40
417	2,000-5,000 (890-2,225)	140-70	637	3,000-6,000 (1,335-2,670)	70-35
427	2,000-5,000 (890-2,225)	140-60	737	3,000-6,000 (1,335-2,670)	65-35
437	1,500-5,000 (668-2,225)	140-60	837	4,000-6,500 (1,780-2,893)	55-30
437	2,000-5,000 (890-2,225)	140-60	837	3,000-6,000 (1,335-2,670)	65-35
437	2,000-5,500 (890-2,448)	120-60	115	1,000-4,000 (445-1,780)	350-80
447	2,000-5,500 (890-2,448)	120-60	135	1,000-4,000 (445-1,780)	350-80

Table 8, cont.
Typical Weights and RPMs for Roller Cone Bits

IADC Code	psi of diameter (dN/mm of dia.)	rpm Range	IADC Code	psi of diameter (dN/mm of dia.)	rpm Range
415	1,000-3,500 (445-1,558)	350-80	127	2,000-5,500 (890-2,448)	150-70
435	1,000-4,000 (445-1,780)	350-80	136	2,000-6,000 (890-2,670)	120-60
			137	2,000-6,000 (890-2,670)	120-60
114	1,500-4,000 (668-1,780)	200-80	216	3,000-7,000 (1,335-3,115)	90-50
115	1,500-4,000 (668-1,780)	200-80	217	3,000-7,000 (1,335-3,115)	90-50
135	1,500-4,000 (668-1,780)	200-80	316	3,000-7,500 (1,335-3,338)	80-50
415	1,000-3,500 (445-1,558)	200-80	317	3,000-7,500 (1,335-3,338)	80-50
435	1,000-4,000 (445-1,780)	200-80	347	3,000-7,500 (1,335-3,338)	80-50
435	1,500-4,000 (668-1,780)	200-80			
445	1,500-4,500 (668-2,005)	200-80	612	3,000-6,000 (1,335-2,670)	80-40
515	1,500-4,500 (668-2,005)	200-80	632	3,000-6,000 (1,335-2,670)	80-40
			732	3,000-6,000 (1,335-2,670)	70-40
126	2,000-5,500 (890-2,448)	150-70	832	3,000-6,000 (1,335-2,670)	70-35

NOTE: It is not recommended that the upper limits of weight and rpm be run simultaneously without consultation with a bit company representative.

Note that in soft formations (IADC code 1), a rotary speed of 300 to 380 rpm is combined with lighter bit weights from 2,000 to 5,000 pounds (lb) or 890 to 2,225 decanewtons (dN) times bit diameter in ft or mm. In hard formations, weights of 3,000 to 7,500 lb (1,335 to 3,338 dN), times bit diameter, are required for the short tooth's crushing and chipping action to overcome the rock's high compressive strength. Steady improvements in the quality of insert bits allow higher speeds in soft formations and higher weights in hard formations than previously possible, but the maximum weight and maximum rpm are not to be run simultaneously.

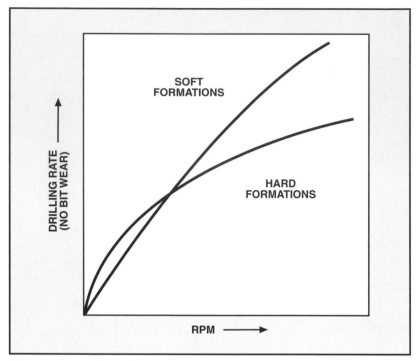

Figure 48. Effect of rotary RPM on drilling rate

For roller cone bits, an increase in rotary speed rate does not correspond directly to an increase in penetration rate—that is, doubling the speed does not double the penetration rate, but more nearly does so in soft formations than in hard (fig. 48).

Natural diamond bits and PDCs are best run at lighter weights and sometimes at higher speeds than roller cone bits. Because of their cost and nature of their cutting action, diamond bits should be run with the lightest weight possible that still achieves a good penetration rate. Weights of 900 to 4,500 lb per in. of bit diameter (15 to 80 dN per mm of bit diameter) are typical, but higher weights may be used in some situations (table 9). Rotary speeds in the range of 100 rpm are normal but a range from 200 to 600 rpm is feasible with downhole mud motors, or even higher with turbine motors. With smooth running, proper hydraulics, and the optimum weight, drilling rates usually increase in direct proportion to rotary speed; however, lower speed improves bit durability.

Table 9
Typical Specifications and Operating Guidelines for
Natural Diamond Bits Run on Downhole Mud Motor

Specifications	8½ in. (215.9 mm)	6½ in. (165.1 mm)	4¾ in. (120.7 mm)
Diamond Size, stones/carat	4-8	4-8	4-8
API Pin Connection, in. (mm)	4½ Reg. (114.3)	3½ Reg. (88.9)	2⅞ Reg. (73.0)
Nominal Gauge Length, in. (mm)	4.5 (114.3)	3.5 (88.9)	1.5 (38.1)
Operating Guidelines (Optimum Ranges)	**8½ in. (215.9 mm)**	**6½ in. (165.1 mm)**	**4¾ in. (120.7 mm)**
Flow Rate, Range, gpm (m³)	325-475 (1.23-1.8)	175-250 (0.66-0.95)	90-150 (0.34-0.56)
HSI, Range, hhp/in² (hhp/cm²)	1.0-3.0 (6.45-19.35)	1.0-3.0 (6.45-19.35)	1.0-3.0 (6.45-19.35)
Weight on Bit, Range, lbs (dN)	9,000-38,000 (4,000-17,000)	6,000-25,000 (2,700-11,000)	5,000-20,000 (2,225-9,000)
Pressure Drop, Range in psi (kPa)	200-600 (1,379-4,137)	200-600 (1,379-4,137)	200-400 (1,379-2,758)
rpm up to	250	400	600

Courtesy of Reed-Hycalog

To summarize—

- Generally, an increase in either weight or rpm increases ROP, provided the right bit is in place and bottomhole cleaning is adequate.
- Other factors play an important role, such as increased likelihood of drill string failure, extra bearing and cutter wear, and hole deviation.
- Normally, higher rotary speeds require lower weight on bit and vice versa.
- In general, diamond bits should be run at lighter weights and higher speeds than roller cone bits.

Special
Considerations

▼
▼
▼

S pecific contract requirements, the rig's capability, and other factors must be considered when calculating the appropriate bit weight and rotary speed.

For example, additional, or heavier, drill collars may be needed to provide the added weight or stiffness. If more and heavier drill collars are required, one question the rig owner must ask is, "Can the derrick safely handle the heavier load?" Other considerations include costs. Drill collars are expensive to buy, or rent, and to maintain. They are hard to handle and require safety clamps and lifting subs that add to the trip time. If ten stands (30 collars) are used, it may require several hours to break out and make up the collars in a round trip.

Another consideration is that deviation and doglegs tend to develop when the bit weight is changed (fig. 49). If the drilling

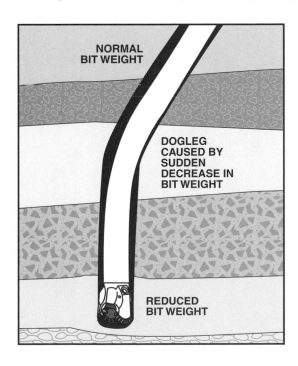

Figure 49. Dogleg produced by reduced weight

contract imposes strict deviation controls, or if dipping formations are present, increasing the bit weight to attain faster penetration may be ill-advised.

The rig owner must also bear in mind that the rig's power system may limit the ability to increase both weight and rpm. Increased weight and rpm increase rotary torque, or resistance to turning. Adding drill collars can increase the torque because of the extra weight and closer hole contact. Higher torque means that more horsepower is required to operate the rotary (fig. 50). Other situations may also require extra horsepower. For instance, long-tooth bits require more rotary power than short-tooth bits drilling the same formation. Doglegs and slanted holes require more horsepower because of the greatly increased wall friction. A dogleg also increases the possibility of drill string failure because of the alternating stresses imposed on the drill string with each rotation (fig. 51).

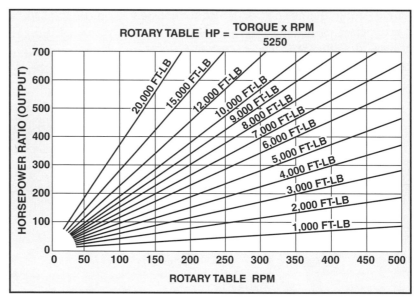

Figure 50.   Rotary drive power required at various torque loads and RPM

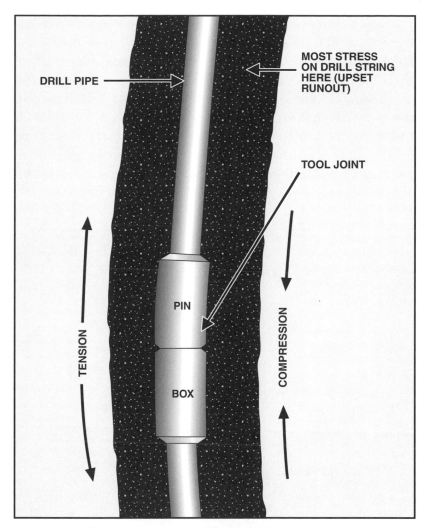

Figure 51. Tension and compression on drill string during rotation

The combination of high rotary speed, torque, and extra weight on the bit creates great stress on the drill string. The drill string is a flexible shaft and smooth running is important. A certain amount of irregularity is unavoidable, but if centrifugal forces are severely unbalanced, wall friction will increase or the bit may wear unevenly. These situations can lead to serious problems such as joint failure, twist-off, bit tooth breakage, or bearing failure.

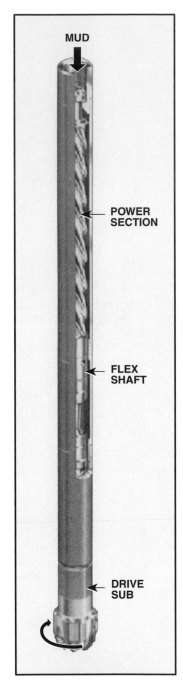

Figure 52. Cut-away drawing of a downhole motor powered by drilling mud (Courtesy of Baker Hughes INTEQ)

Slant and horizontal drilling use a downhole motor to overcome the wall friction problem (torque) and to avert drill string failure. A downhole motor is a length of strong pipe inside of which there is a spiral device called the stator/rotor or, in turbines, a series of blades. The downhole motor or turbine is attached directly to the bit. The pump pressure forces the mud past the stator or turbine device, turning it along with the bit. The bit rotates but the drill string does not (figs. 52, 53). Downhole motors require more pump capacity than regular rotary drilling in order to rotate the bit at extremely high speeds (up to 2,000 rpm).

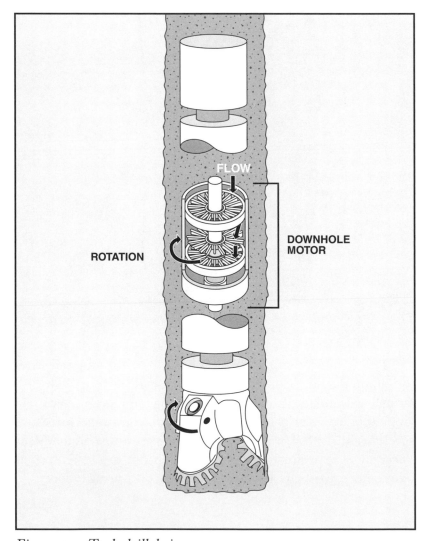

Figure 53. Turbodrill design

To initiate the turn in a horizontal hole, the drilling assembly usually includes a bent sub or a bent housing placed above the downhole motor (fig. 54). The turn radius may be short or long (fig. 55).

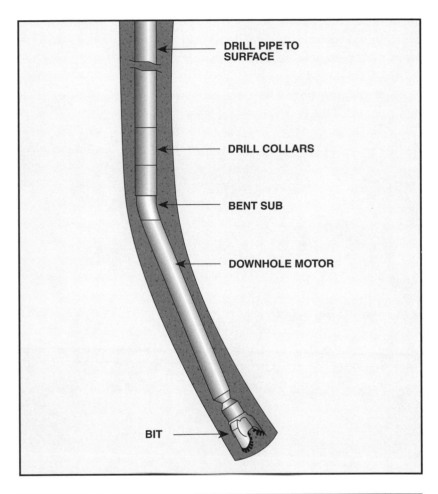

Figure 54. Drilling assembly using a downhole motor and a bent sub

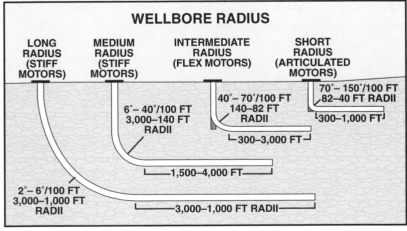

Figure 55. Horizontal turn radii (Courtesy of Baker Hughes INTEQ)

The turn is steered by use of sophisticated downhole instruments tied into the rig site computers. With bent subs, the drill string must be pulled and the bent sub changed in order to change the angle of inclination. In the newest adjustable kick-off (AKO) motors the bit can be steered without pulling the drill string to reset the angle or direction of inclination (figs. 56 and 57). Downhole mud motors may be stiff, flexible or articulated.

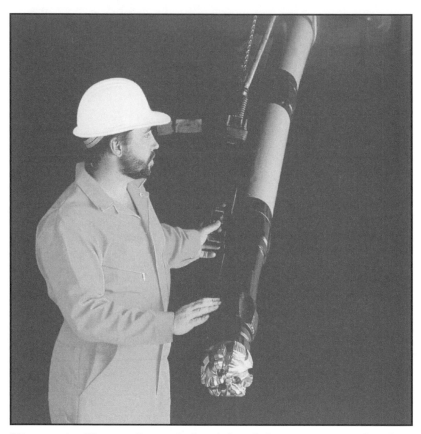

Figure 56. Short radius articulated downhole motor equipped with PDC bit (Courtesy of Baker Hughes INTEQ)

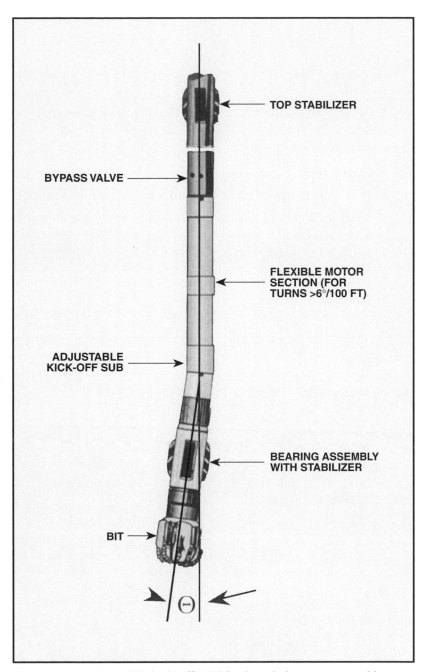

Figure 57. Adjustable kick-off (AKO) downhole motor assembly can be used for drilling vertical to short-radius horizontal holes. (Courtesy of Baker Hughes INTEQ)

To summarize—

Special considerations when determining appropriate bit weight and rotary speed:

- More drill collars are required to put additional weight on bit
- Can derrick handle additional loads?
- Drill collars are expensive
- Drill collars are more difficult to handle than drill pipe
- Doglegs and hole deviation can occur when weight on bit is changed
- Rig power system limitations
- Drill string stresses
- Downhole motors

Rate of Penetration Control

Variations in the drilling rate are normal. Changes can indicate bit wear, change in formation, weight, rotary speed, or hydraulics. The driller must evaluate all these possibilities before taking corrective action.

An unworn bit matched to the right formation, and properly run, will drill faster than a worn bit or a bit not matched to the formation. A driller can use this fact to determine the optimum weight to maintain by measuring how much time is required to drill a certain distance—1 ft, 10 ft (1 m, 10 m), or a kelly length. This time measurement is then converted to ft/h or m/h (the usual basis for comparing bit performance) or minutes (min) per ft or m. The rig's drilling rate recorder (Geolograph™) allows a quick visual evaluation of the drilling rate (fig. 58). Computers tied directly to the rig instruments are increasingly being used to collect such data. If automatic devices are not available, the driller can determine drilling rate by marking the kelly, maintaining a constant speed and bit weight, then measure the time it takes to drill one foot. This provides a rough estimate of bit performance under that particular weight and speed.

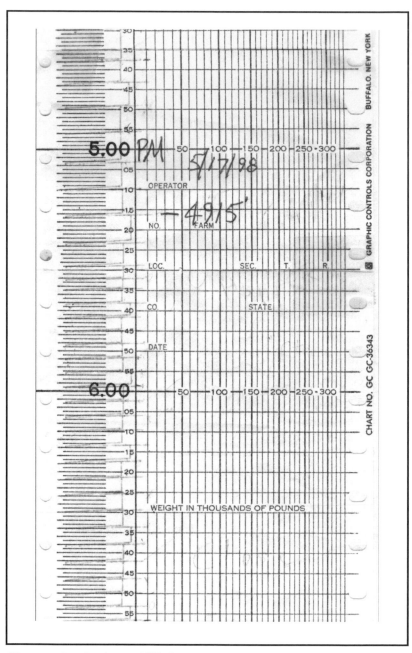

Figure 58.   Geolograph™ *chart.   Note drilling break at 4,915 feet where average ROP decreases from average 6.5 min/ft to 3.5 min/ft. (Courtesy of GLB Explorer)*

A common method to determine the optimum weight and rotary speed is the *drill-off* technique. A drill-off test uses the fact that a drill string is elastic. Its length varies with the tension on the string. As more drill string weight is suspended from the crown block, there is less weight on the bit. Maintaining a constant rotary speed and circulation, a given amount of weight is imposed (*slacked off*) on the bit and the brake tied down. The time it takes for the bit to drill off some increment (usually 2,000 to 5,000 lb or 1,000 to 2,000 dN) of the applied weight is recorded. As the weight is drilled off, more of the drill string weight is suspended from the crown block and less is applied to the bit. The time it takes to drill off the next increment of weight is then recorded, and so on until the test is completed (table 10). The time it takes to drill off each increment of bit weight varies. In the test recorded on table 10, the best time was 25 seconds using a weight of 60,000 lb (26,700 dN). This weight is the optimum bit weight for the existing conditions at the time of the test.

Table 10
Sample Drill-Off Test in Sandy Shale

Weight Interval, 1,000 lb (dN)	Drill-Off Time, seconds
70–65 (31,150–28,925)	26
65–60 (28,295–27,000)	26
60–55 (27,000–24,475)	25 (Best Run)
55–50 (24,475–22,250)	28
50–45 (22,250–20,025)	31
45–40 (20,025–17,800)	34
40–35 (17,800–15,575)	36
35–30 (15,575–13,350)	52
30–25 (13,350–11,125)	70

By making a pick-up and slack-off chart the time (in seconds) to drill off a certain amount of weight can be converted into ft (m) drilled per hour. Figure 59 is based on the drill-off test of table 10. The chart shows that the kelly moved 0.06 ft (0.018 m) for each 1,000 lb (445 dN) drilled off. Using the best run of 25 seconds to drill off 5,000 lb (2,225 dN), the hourly drilling rate can be calculated as follows.

Drilling rate = 3,600 seconds × 0.06 × weight drilled off in 1,000-lb increments ÷ 25 seconds = 3,600 × 0.06 × = 43 ft/h (13 m/h).

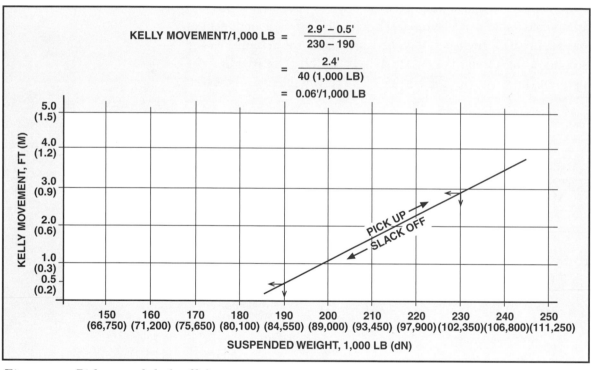

Figure 59. Pick-up and slack-off chart

Drill-off tests can be used to determine the effectiveness of other drilling variables. The weight and hydraulics can be held constant and the rotary speed varied, for instance. Optimum hydraulics can likewise be determined. Drill-off tests should be repeated periodically or when there is a change in formation or other operating conditions.

To summarize—

Drill-off technique to determine optimum weight on bit and rotary speed

- Maintain constant rotary speed and circulation
- Slack off a given amount of weight on bit
- Record time it takes for bit to drill off a given increment of weight (e.g., 2,000 to 5,000 lb or 1,000 to 2,000 dN); repeat several times
- From tests, determine least amount of time to drill off weight

---

Drilling Mud

Drilling mud properties impact the penetration rate by performing functions vital to cost-effective drilling (fig. 60). The basic functions of drilling mud are to:

1. clean the bit teeth and the bottom of the hole;

2. transport formation cuttings to the surface;

3. prevent formation fluids from entering the wellbore causing a kick or blowout;

4. protect and support the walls of the wellbore;

5. cool and lubricate the bit and drill string;

6. provide hydraulic power for downhole motors or turbines (see figs. 52 and 53); and

7. help detect the presence of oil, gas, or saltwater in formations.

Drilling mud contains three types of material—one liquid and two solid. The liquid component may be water (water base) or oil (oil base), or a mixture of both.

One type of solid is reactive with liquid. The main reactive solids in most drilling muds are clays. Clays swell in water and thicken the mud. The other type of solids is nonreactive which means they do not react with the liquid phase of the mud. Nonreactive solids include formation cuttings of all sizes. Solids are generally undesirable because they add weight to the mud and are abrasive to equipment. One common nonreactive solid is barite, which is purposely added to the mud to increase the weight as needed to control formation pressure.

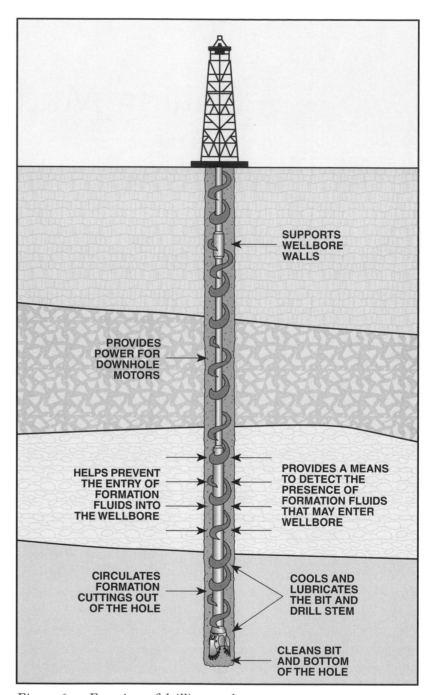

Figure 60. Functions of drilling mud

The mud program is not the same for every well. In the course of drilling a well, the mud can be changed to deal with changes in formation and mechanical factors that affect the drilling rate. Mud companies have formulated sophisticated compositions of mud that allow the driller more control under different drilling conditions. Mud records from nearby wells provide a good reference for cost control and well planning (fig. 61). Mud properties with the greatest effect on penetration rate are—

- density,
- viscosity,
- solids content,
- fluid loss, and
- oil content.

Density is the weight of a mud. Mud weight, or mud density, is usually expressed as a unit of weight per unit of volume, such as pounds per gallon (ppg), pounds per cubic foot (pcf), or kilograms per cubic metre (kg/m³). Density directly affects the hydrostatic pressure of the mud column—the pressure the column of mud exerts at a given depth. A heavy mud exerts more hydrostatic pressure at a given depth than a light mud. For example, a mud with a density of 12 ppg (1,437.8 kg/m³) exerts more hydrostatic pressure at 1,000 ft (300 m) than a 10-ppg (1,198.2-kg/m³) mud. Crew members can measure mud weight and calculate its hydrostatic pressure at a given depth using a simple formula. The formula is—

$$P_h = D \times W_m \times C$$

where
P_h = hydrostatic pressure
D = depth
W_m = mud weight
C = a constant

The value of C, the constant, depends on the units used to express the mud weight. If the mud weight is in ppg, the constant is 0.052. If the mud weight is in pcf, the constant is 0.00694. If the mud weight is in kilopascals (kPa), the constant is 0.0098. Thus, hydrostatic pressure in pounds per square inch (psi) where mud weight is measured in ppg is—

$$P_h \text{ (psi)} = D \text{ (ft)} \times W_m \text{ (ppg)} \times 0.052.$$

For example, determine the hydrostatic pressure at the bottom of a hole 5,750 ft deep that is full of a mud whose weight is 11.4 ppg.

Mud Characteristics That Affect ROP

Density

73

WATER-BASED MUD REPORT No. 14

Date	8/22/99	Depth/TVD	9052 ft / 9052 ft
Spud Date	8/ 9/99	Mud Type	F.W.N.D
Water Depth		Activity	DRILLING

Operator :
Report For :
Well Name :
Contractor : XYZ
Report For : John Doe

Field/Area :
Description :
Location : GRADY CO. OK
Well No. : 370

DRILLING ASSEMBLY

Bit Size 7.88 in W/M 535P	
Nozzles 14 /14 /14 1/32"	
Drill Pipe Size 4.5 in	Length 8474 ft
Drill Pipe Size in	Length ft
Drill Collar Size 6.25 in	Length 578 ft

CASING

Surface	8.625in @1010ft (1010TVD)
Intermediate	4.5in @9150ft (9150TVD)
Intermediate	
Production or Liner	

MUD VOLUME (bbl)

Hole	481.1
Active Pits	479.9
Total Circulating Vol	961
In Storage	

CIRCULATION DATA

Pump Make	IDECO D-550	IDECO D-700
Pump Size	6 X 15.in	6 X 16.in
Pump Cap	6.035 gal/stk	6.309 gal/stk
Pump stk/min	58@90%	
Flow Rate	350 gal/min	
Bottoms Up	42.9 min	2488 stk
Total Circ Time	115.3 min	6689 stk
Circulating Pressure	1650 psi	

MUD PROPERTIES

Sample From		F.L.@0800
Flow Line Temp	°F	115
Depth/TVD	ft	9052/9052
Mud Weight	lb/gal	9.45@115°F
Funnel Viscosity	s/qt	43
Rheology Temp	°F	100
R600/R300		35/21
R200/R100		
R6/R3		
PV	cP	14
YP	lb/100 ft²	7
10s/10m/30m Gel	lb/100 ft²	4/17/27
API Fluid Loss	cc/30 min	12.0
HTHP FL Temp	cc/30 min	
Cake API/HT	1/32"	2/
Solids	% Vol	8.3
Oil/Water	% Vol	0/91.7
Sand	% Vol	0.25
MBT	lb/bbl	25
pH		10.3
Alkal Mud (Pm)		1.1
Pf/Mf		.15/.45
Chlorides	mg/l	1800
Hardness Ca		80
LCM	PPB	-----

PRODUCTS USED LAST 24 HRS

Products	Size	Amt
CAUSTIC SODA	50 LB BG	4
DRAYAGE	1 EA	173
M-I GEL	100 LB BG	140
SP 101	50 LB BG	1

SOLIDS EQUIP

	Size	Hr
Shale Shaker	100	24
Desander	8 X 2	18

MUD PROPERTY SPECS

Weight	9.4-9.5
Viscosity	48-50
Filtrate	-15 CC

REMARKS AND TREATMENT

GEL>>>>>3-10 MIN/SX FOR 48-50 VIS

SODA ASH>*DAY* TOUR ADD 3 SX @ 30 MIN>SX

CAUSTIC>>>NONE

WATER>>ONLY IF NEEDED TO HOLD WT @ 9.5

REMARKS

VISCOSITY>>>48- 50 SEC>QT
MUD WT>>>>>9.4-9.5 PPG

SHORT TRIP BEFORE TOH TO LOG

TIME DISTR Last 24 Hrs

Rig Up/Service	
Drilling	
Tripping	
B.O.P. NU	
B.O.P. Testing	
Cementing	
Condition Hole	
Condition Mud	
Coring	
Dev. Survey	

MUD VOL ACCTG (bbl)

Oil Added	
Water Added	74
Mud Received	0
Mud Disposed	0
Shakers Loss	0
Evaporation Loss	0
Centrifuge Loss	0
Formation Loss	0
Left in Hole	0
Other	0

SOLIDS ANALYSIS (%/lb/bbl)

NaCl	.1/ 1.
KCl	./ .
Low Gravity	8.3/ 75.4
Bentonite	2.1/ 18.7
Drill Solids	6.2/ 56.6
Weight Material	NA/ NA
Chemical Conc	- / .
Inert/React	2.0141
Average SG	2.6
Carb/BiCarb (m mole/L)	2.8/ .7

MUD RHEOLOGY & HYDRAULICS

np/na Values	0.737/0.737
kp/ka (lb•s^n/100ft²)	0.226/0.226
Bit Loss (psi / %)	524 / 31.8
Bit HHP (hhp / HSI)	107 / 2.2
Bit Jet Vel (ft/sec)	249
Annular Vel DP (ft/min)	205.39
Annular Vel DC (ft/min)	373.73
Crit Vel DP (ft/min)	167
Crit Vel DC (ft/min)	256
ECD @ 9052 (lb/gal)	9.75

M-I ENGR / PHONE	RIG PHONE	WAREHOUSE PHONE	DAILY COST	CUMULATIVE COST
JOE SIMMONS (405) 620-7098	(580) 571-7021	(590) 223-1822	$ 864.80	$ 5,098.46

Figure 61. Daily drilling mud report (water base mud) from previously drilled well. Computerized report as printed out at rig site by mud engineer. (Courtesy of M-I L.L.C.)

Thus—

$$P_h = 5{,}750 \text{ ft} \times 11.4 \times 0.052$$
$$= 65{,}550 \times 0.052$$
$$P_h = 3{,}409 \text{ psi.}$$

Hydrostatic pressure in psi where mud weight is measured in pcf is—

$$P_h \text{ (psi)} = D \text{ (ft)} \times W_m \text{ (pcf)} \times 0.00694.$$

For example, determine the hydrostatic pressure at the bottom of a hole 5,750 ft deep that is full of mud whose weight is 85.3 pcf. Thus—

$$P_h = 5{,}750 \text{ ft} \times 85.3 \times 0.00694$$
$$= 490{,}475 \times 0.00694$$
$$P_h = 3{,}404 \text{ psi.}$$

Hydrostatic pressure in kPa where mud weight is measured in kg/m³ is—

$$P_h \text{ (kPa)} = D \text{ (m)} \times W_m \text{ (kg/m}^3) \times 0.0098.$$

For example, determine the hydrostatic pressure at the bottom of a hole 1,750 m deep that is full of mud whose weight is 1,365.9 kg/m³. Thus—

$$P_h = 1{,}750 \times 1{,}365.9 \times 0.0098$$
$$= 2{,}390{,}325 \times 0.0098$$
$$P_h = 23{,}425 \text{ kPa.}$$

Table 11 is a mud-weight density conversion table using common mud-density measurements. When the mud weight is in ppg, crew members can use table 12 to make mud weight adjustments using barite or water to attain a specific hydrostatic pressure.

Lightweight muds (less than 10 ppg or 1,198.2 kg/m³) exert less pressure on the bottom of the hole and allow cuttings to be removed efficiently with lower weight and rotary speed. In effect, the rock drills more easily, provided the circulation system is properly maintained. Drilling with lightweight mud with its lower hydrostatic pressure can, however, increase the risk of a kick.

On the other hand, if mud density is too high, high differential pressure exists between the mud column and the formation. Put another way, the mud's hydrostatic pressure is higher than the formation pressure. Hydrostatic pressure higher than formation pressure creates a chip hold-down effect that tends to hold the cuttings on the bottom of the hole. Unless mechanical energy is increased, a drop in drilling rate occurs because the bit will be drilling the same material over and over.

Table 11
Mud Density Conversion Table

pounds per gallon (lb/gal)	pounds per cubic foot (lb/ft³)	grams per cubic centimetre (g/cm³)*	kilograms per cubic metre (kg/m³)	pounds per gallon (lb/gal)	pounds per cubic foot (lb/ft³)	grams per cubic centimetre (g/cm³)	kilograms per cubic metre (kg/m³)
6.5	48.6	0.78	780	16.0	119.7	1.92	1,920
7.0	52.4	0.84	840	16.5	123.4	1.98	1,980
7.5	56.1	0.90	900	17.0	127.2	2.04	2,040
8.0	59.8	0.96	960	17.5	130.9	2.10	2,100
8.3	62.3	1.00	1,000	18.0	134.6	2.16	2,160
8.5	63.6	1.02	1,020	18.5	138.4	2.22	2,220
9.0	67.3	1.08	1,080	19.0	142.1	2.28	2,280
9.5	71.1	1.14	1,140	19.5	145.9	2.34	2,340
10.0	74.8	1.20	1,200	20.0	149.6	2.40	2,400
10.5	78.5	1.26	1,260	20.5	153.3	2.46	2,460
11.0	82.3	1.32	1,320	21.0	157.1	2.52	2,520
11.5	86.0	1.38	1,380	21.5	160.8	2.58	2,580
12.0	89.8	1.44	1,440	22.0	164.6	2.64	2,640
12.5	93.5	1.50	1,500	22.5	168.3	2.70	2,700
13.0	97.2	1.56	1,560	23.0	172.1	2.76	2,760
13.5	101.0	1.62	1,620	23.5	175.8	2.82	2,820
14.0	104.7	1.68	1,680	24.0	179.5	2.88	2,880
14.5	108.5	1.74	1,740				
15.0	112.5	1.80	1,800				
15.5	115.9	1.86	1,860				

*Same as specific gravity (sg).

$$\text{Mud Gradient, psi/ft} = \frac{\text{lb/ft}^3}{144}, \quad \frac{\text{lb/gal}}{19.24}, \quad \text{or} \quad \frac{\text{kg/m}^3}{2,309}$$

$$\text{Density} = \text{g/cm}^3 = \frac{\text{lb/ft}^3}{62.3} = \frac{\text{lb/gal}}{8.345}$$

Table 12
Mud-Weight Adjustment with Barite or Water

Initial Mud Weight (ppg)	Desired Mud Weight (ppg)																	
	9.5	10.0	10.5	11.0	11.5	12.0	12.5	13.0	13.5	14.0	14.5	15.0	15.5	16.0	16.5	17.0	17.5	18.0
9	29	59	90	123	156	192	229	268	308	350	395	442	490	542	596	653	714	778
9.5		29	60	92	125	160	196	234	273	315	359	405	452	503	557	612	672	735
10	43		30	61	93	128	164	201	239	280	323	368	414	464	516	571	630	691
10.5	85	30		31	62	96	131	167	205	245	287	331	376	426	479	531	588	648
11	128	60	23		31	64	98	134	171	210	251	294	339	387	437	490	546	605
11.5	171	90	46	19		32	66	101	137	175	215	258	301	348	397	449	504	562
12	214	120	69	37	16		33	67	103	140	179	221	263	310	357	408	462	518
12.5	256	150	92	56	32	14		34	68	105	144	184	226	271	318	367	420	475
13	299	180	115	75	48	27	12		34	70	108	147	188	232	278	327	378	432
13.5	342	210	138	94	63	41	24	11		35	72	111	150	194	238	286	336	389
14	385	240	161	112	76	54	36	21	10		36	74	113	155	199	245	294	345
14.5	427	270	185	131	95	68	48	32	19	9		37	75	116	159	204	252	303
15	470	300	208	150	110	82	60	43	29	18	8		37	77	119	163	210	259
15.5	513	330	231	169	126	95	72	54	39	26	16	8		39	79	122	168	216
16	556	360	254	187	142	109	84	64	48	35	24	15	7		40	81	126	172
16.5	598	390	277	206	158	123	96	75	58	44	32	23	14	7		41	84	129
17	641	420	300	225	174	136	108	86	68	53	40	30	21	13	6		42	86
17.5	684	450	323	244	189	150	120	96	77	62	49	38	28	20	12	6		43
18	726	480	346	262	205	163	132	107	87	71	57	45	35	26	18	12	5	

The lower left half of this table shows the number of barrels of water that must be added to 100 bbl of mud to produce desired *weight reductions*. To use this portion of the table, locate the initial mud weight in the vertical column at the left, then locate the desired mud weight in the upper horizontal row. The number of barrels of water to be added per 100 bbl of mud is read directly across from the initial weight and directly below the desired mud weight. For example, to reduce an 11 ppg mud to a 9.5 ppg mud, 128 bbl of water must be added for every 100 bbl of mud in the system.

The upper right half of this table shows the number of sacks of barite that must be added to 100 bbl of mud to produce desired *weight increases*. To use this portion of the table, locate the initial mud weight in the vertical column to the left, then locate the desired mud weight in the upper horizontal row. The number of sacks of barite to be added per 100 bbl of mud is read directly across from the initial weight and directly below the desired mud weight. For example, to raise an 11 ppg mud to 14.5 ppg, 251 sacks of barite must be added per 100 bbl of mud in the system.

Using low-weight muds when possible is a definite cost saver (fig. 62). Savings include less rotating time and trip time, lower bit costs, fewer occurrences of lost circulation, stuck pipe, and other work interruptions.

Sometimes a heavy mud of 16 to 18 ppg (1,917 to 2,157 kg/m³) is required to control overpressured formations. If sufficient hydrostatic pressure is not maintained, the well may kick. Keep in mind, however, that hydrostatic pressure should not exceed formation pressure by too much. A hydrostatic pressure that is too high may fracture the formation or enlarge existing fractures and circulation may be lost. Drilling with the proper mud weight and carefully maintaining other mud characteristics is the best way to prevent lost circulation and kicks.

If an overpressured formation is anticipated, an intermediate string of casing (protection pipe) may be run. Intermediate casing protects normally pressured shallow formations from excessive hydrostatic pressure and prevents lost circulation within up-hole zones as mud weights are increased.

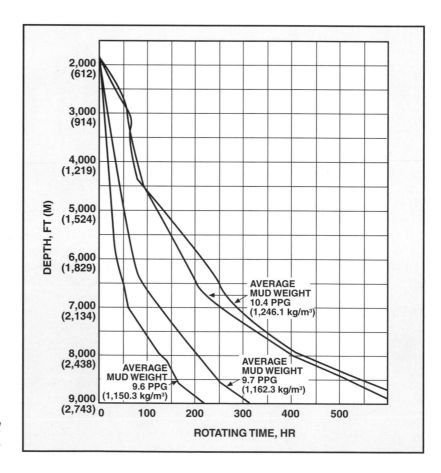

Figure 62. Effect of mud weight on drilling rate

Viscosity is a measure of a fluid's resistance to flow. To increase the viscosity of natural mud, a reactive clay that swells in water is added to the system. Bentonite clay is the most common viscosity additive but chemical polymers are often used. A thick, or viscous, mud is more difficult to pump than a thin, less viscous mud. The more viscous a mud is, the more pressure that is required to pump it. Thus, as mud viscosity increases, circulating (pump) pressure decreases. A decrease in circulation pressure decreases the bit hydraulic horsepower (bhhp). Loss of bhhp means bottomhole cleaning is less efficient and the penetration rate slows because the thick mud tends to hold formation chips on the bottom.

Viscosity

Gel strength, the ability of a mud to keep cuttings from settling when circulation is stopped, is a function of viscosity. Gel strength, as well as viscosity, should be closely monitored to assure a clean hole.

Sometimes a quantity of high-viscosity mud is circulated through the system for hole-cleaning purposes. Known as a *high-viscosity sweep*, this operation clears the hole of junk before setting casing or running a diamond bit.

If the viscosity is too high, water can be added to thin the mud but water will also lower the mud density. When water lowers the density too much, weighting material must be added to the mud to bring the density back to the proper level. Mud companies supply chemical polymers that lower viscosity without reducing density.

The proportion, size, and type of solids in mud greatly affect its properties. Mud engineers control these properties by installing equipment to remove drilled solids and supervising the amount, timing, and types of solids added to the mud. Figure 63 illustrates the effect of mud solids content on the drilling rate.

Solids Content

Low-solids mud is maintained in drilling situations where mud weights are no greater than 10 ppg (1,198.2 kg/m³) and circulation rates are kept high enough to lift cuttings out of the hole. Small particles of weighting material (barite) can retard penetration by plugging fractures induced by the bit cutters. This tends to hold the chips in place and inhibits cuttings removal, thus decreasing the penetration rate.

There are a number of ways to maintain low-solids content in mud. Circulating mud through the reserve pit allows fine solids to settle out. Devices like desilters, desanders, and centrifuges (for heavy muds) mechanically control solids content. Chemicals are also used to cause fine particles to aggregate and settle out as larger particles.

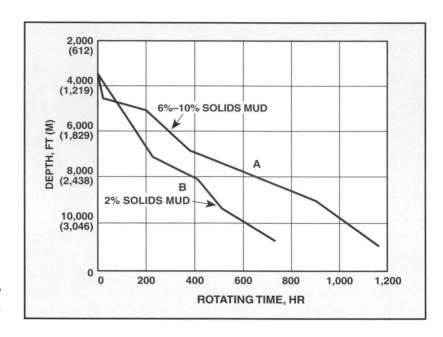

*Figure 63. Effect of mud
solids on drilling rate*

Water is the ultimate low-solids drilling fluid, enabling even faster penetration rates than lightweight mud. Where water supply and downhole conditions permit, so-called clear water drilling is employed, often to depths of several thousand feet (metres).

Fluid Loss   The hydrostatic pressure exerted by the mud column forces some of the liquid component of the mud into the pore spaces of the rocks. Because water-base muds are so often used, the liquid component of the mud is often water. Consequently, fluid loss is often termed water loss. Because other liquids besides water make up mud, fluid loss is also termed filtration loss. As liquid is lost to the formation, a layer of particles called wall cake (filter cake) builds up on the wall of the hole. Three benefits of wall cake are: 1) it protects the borehole walls; 2) it helps prevent the entry of formation fluids into the borehole; and 3) it seals the borehole walls which prevents the loss of whole mud into the formation.

Excessive water loss, however, causes serious problems. A thick filter cake buildup can cause a stuck drill string (fig. 64). The amount of formation wetting is apparently related to the water loss of the drilling mud. Wetting of a shale formation causes sloughing and caving to occur that may, again, lead to stuck drill pipe. Excessive water loss may also hinder wireline log interpretation and cause well completion problems.

An oil-emulsion mud is a water-base mud with a little oil mixed in. Oil and water do not mix, so an emulsifier is added to the mud,

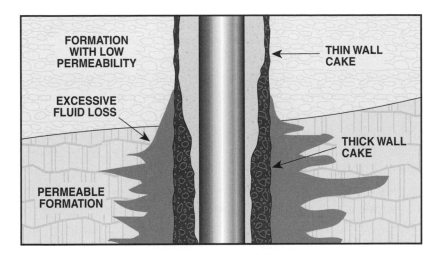

Figure 64. Buildup of filter (mud) cake on bore-hole walls. Solid particles in the drilling mud plaster the wall of the hole forming an impermeable barrier.

along with the oil, to disperse the oil and prevent it from separating in the water-base system. The addition of oil to the mud can affect the penetration rate in certain formations, particularly where high temperatures, sloughing shale, or drill pipe sticking may be expected. The oil in the mud lowers friction and torque, thus reducing horsepower requirements. The lubricating ability of oil increases bit life. Oil in the mud also keeps the bit from balling in hydrating clays and shales.

Oil-base mud has oil, usually diesel, as the liquid phase (instead of water). Some water is always present but kept at 5 percent or lower. Oil-base mud must contain an emulsifier for the water. Because of environmental concerns when using oil-base muds, mud companies have developed synthetic oil-base muds. Instead of using diesel or other aromatic mineral oils, synthetic muds have vegetable oils or nontoxic mineral oils. Whether synthetic or not, oil-base mud is much more expensive than water-base mud, so its use is usually restricted to specific, difficult drilling situations. Such situations include—

- water soluble formations,
- protecting producing formations,
- severe corrosion problems,
- deep, high-temperature holes,
- coring, and
- unsticking of downhole equipment.

Invert-emulsion mud is another type of oil-base mud in which the water is spread out (dispersed) in the oil. The water content may be as high as 10 to 60 percent. The emulsified water lessens the concentration of mixed soap emulsifiers and asphaltic materials required to carry the cuttings and weighting material. The use of invert-emulsion mud is similar to that of oil-base mud.

Mud costs are a significant part of the total drilling cost of a well and the mud program is a critical part of well planning. Mud records (or recaps) from nearby wells (see fig. 61) are usually available and allow operators to forecast mud costs accordingly. Special equipment, storage facilities, additives, and chemicals all represent added costs, as does the time used to mix, treat, maintain, and circulate the mud. A low-density clay-water mud system is relatively inexpensive because water is the cheapest material that can be added to drilling mud. High-density mud is more expensive to prepare and maintain than low-density mud because barite is the most expensive mud additive. Regardless of the type of mud used, the lowest total mud costs are obtained by using the minimum mud weight and volume that meet the demands of the drilling operation

Many mud additives and chemicals are classed as hazardous materials. It is therefore imperative that the crew is trained in the safe and efficient handling of the mud system.

To summarize—

Drilling mud
- Cleans the bit teeth and bottom of hole
- Transports cuttings to the surface
- Prevents formation fluids from entering the hole
- Protects and supports the walls of the wellbore
- Provides hydraulic power for downhole motors
- Helps detect oil, gas, or salt water in formations

Three phases of drilling mud
- Liquid, usually water, oil, or a mixture of water and oil
- Reactive solids, such as bentonite or other clays
- Nonreactive solids, such as barite or some drilled solids

Mud properties that affect ROP
- Density (weight)
- Viscosity (thickness)
- Solids content
- Fluid loss
- Oil content

Air or Gas Drilling

In certain areas, use of air or gas rather than mud as the circulating fluid permits much lower-cost drilling. Generally, air or gas drilling is used in areas where the subsurface formations are older, hard rocks, and where soft, sloughing shale is not a problem. Moreover, formation water production cannot exceed 50 barrels per hour (bbl/h) or 8 m^3/h. These requirements virtually eliminate the Gulf Coast and offshore for the application of air or gas drilling.

Where conditions are favorable, air or gas drilling can offer advantages—

1. Faster penetration rates than drilling with mud because air or gas is the least-dense circulation medium available. An air or gas column does not create a hold-down effect because it holds very little hydrostatic pressure on the rock. Air removes the cuttings instantly so there is no redrilling of loose chips and less dulling of the bit.

2. Air or gas does an excellent job of cooling the bit. As air or gas leaves the bit, it expands and cools. Effective cooling reduces bit bearing wear, which means that bits last longer.

3. Formation changes are instantly recognized by changes at the blooey line—the line out of which the air or gas and cuttings blow to the surface. Rock type can be easily identified.

4. Shows of water, gas, or oil are quickly evident. Formation evaluation is thus accomplished while drilling, negating the need for expensive testing operations.

Significant disadvantages to the use of air or gas as a circulating fluid exist, however. Air or gas cannot exert enough pressure on the walls to prevent formation fluids from entering the borehole. Most wells encounter water-bearing formations at some point during drilling. A limited ability to handle large volumes of water is a serious disadvantage when drilling with air. A second, and near equal disadvantage, is that air or gas cannot prevent the formation, especially shale, from sloughing into the borehole and sticking the drill pipe. If water is present, the problem of sloughing or swelling shale is made worse.

Two other problems must be addressed when drilling with air: 1) the ever-present danger of explosion and fire at the surface, or downhole explosions that cause lost hole or stuck pipe. A spark- and explosion-proof electrical system is required and fire preven- tion rules must be strictly enforced; 2) corrosion of the drill string can occur. Chemicals are available to combat this problem but the added cost must be considered.

Air Drilling Equipment

In air or gas drilling, the circulating fluid does not circulate continu- ously as does drilling mud. Skid-mounted air compressors furnish the high-pressure air supply for a regular rotary rig equipped for air drilling (fig. 65). The air makes a single trip from the air compres- sors down the drill stem and returns through the annulus to the surface where it is vented to the atmosphere through a blooey line.

Figure 65. Skid-mounted air compressors for air drilling

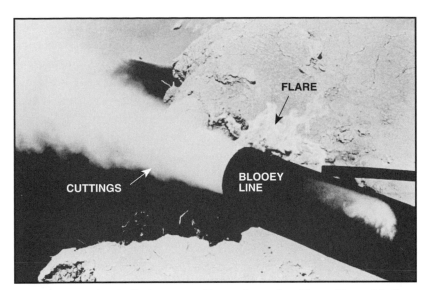

Figure 66. Air drilled cuttings blown through a blooey line with flare set to ignite any gas encountered

To reduce the fire hazard, the blooey line is extended 200 feet or more downwind from the rig where a flare is installed to ignite any formation or drilling gas that may be vented (fig. 66).

In drilling with gas, the gas supply is run directly (through a meter) from a nearby high-pressure gas line. Gas is often used for drilling through producing oil zones.

Air compressors may be single-stage or multiple-stage units. Their capability may range from 50 to 3,000 ft³/min (1.4 to 84 m³/min), with pressures up to 3,500 psi (24,133 kPa). Booster compressors can be hooked up if water is encountered and foam or mist drilling is necessary. Foam or mist drilling involves injecting a surface active agent, a surfactant, into the air or gas stream. The surfactant makes it possible to continue drilling when the borehole encounters relatively small amounts of water because it suspends the drilled formation solids in the water and keeps them from balling up.

Other specialized equipment required for air drilling includes chemical treatment equipment to combat corrosion, mist pumps for injecting foaming agents and fluids, rotating head blowout preventers, and air bits with heavy shanks.

An air hammer and bit is a special air-drilling tool (fig. 67) that may be used to achieve good penetration rates in certain hard rock formations. Air operates the hammer and circulates cuttings. A constant weight and slow rotary speed is applied to the bit while the hammer tool directs repeated blows to the bit, much like holding a chisel firmly against an object and striking the chisel repeatedly with a sledge. Hammer bits may be fitted with tungsten carbide or diamond enhanced inserts similar to cone bits.

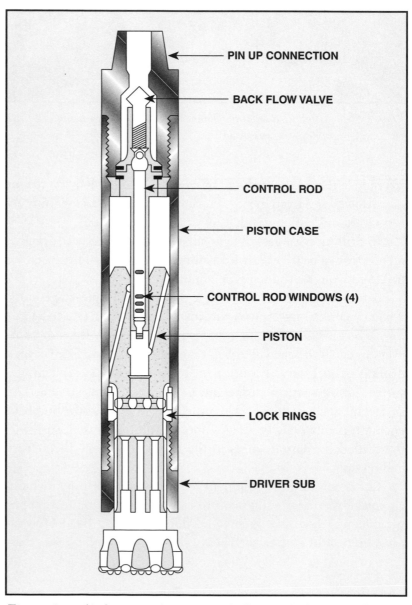

Figure 67. Air hammer (percussion) drilling tool (Courtesy of Smith Bits)

A hammer tool and bit is especially useful where hard rocks are at the surface and it is difficult to apply sufficient weight to the bit. When the hammer tool cuts enough hole, or water overcomes the air supply, drilling may then be changed over to regular rotary methods. In hard rock areas like the Appalachian and Arkoma Basins, the hammer tool may be used to drill the entire hole.

To convert a rotary rig from mud to air drilling, the following procedure is commonly used (fig. 68):

1. Rig up for using mud as the circulating fluid.
2. Install compressors or a connection to a high-pressure gas line.
3. Hook up blowout preventers (BOPs) and bypass lines.
4. Put safety precautions into effect to minimize the fire danger.
5. Install a rotating BOP to make a seal around the kelly or drill pipe. The BOP prevents air or gas from escaping around the drill string but allows the drill string to rotate (fig. 69).
6. Connect a blooey line to vent the air, gas, and cuttings a safe distance downwind from the rig.

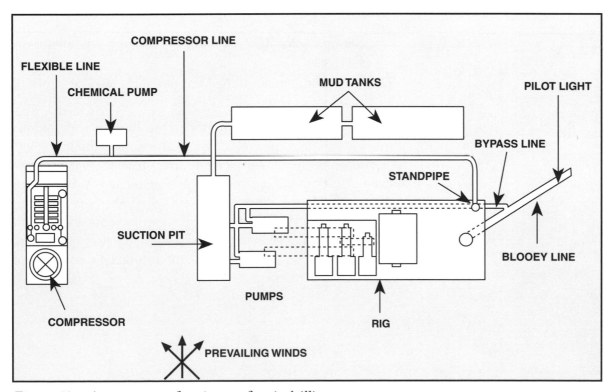

Figure 68. Arrangement of equipment for air drilling

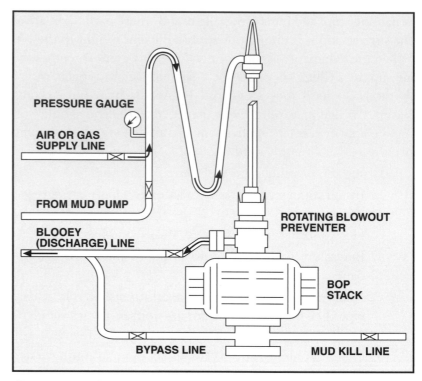

Figure 69. Blowout preventers and lines for air drilling

7. Set up a pilot light, or flare, to burn any gas leaving the blooey line.
8. Install equipment needed for special chemical and mist or foam treatment.

The operator will often rent an entire air drilling package and the contractor will then convert the rig to air or gas drilling.

Volume, pressure, and velocity are variables that must be controlled in air or gas drilling and each are related. A decrease or increase in one factor affects the other two. Most of the problems encountered in air drilling are caused by insufficient volume of air to create the annular velocity needed to clean the hole of cuttings. This condition may occur because of insufficient compressor capacity or hole enlargement.

Table 13

Approximate Rate of Circulation in ft³/min in 8¾-inch hole (m³/min in 222-mm hole) with 4½-inch (114-mm) Drill Pipe with Volumes to Produce Lifting Power Equivalent to a Velocity of 3,000 ft³/min (915 m³/min)

ROP	30 ft³/min (9 m³/min)	90 ft³/min (27 m³/min)	30 ft³/min (9 m³/min)	90 ft³/min (27 m³/min)
Well Depth	Compressed Air		Natural Gas	
	Circulation Rates			
2,000 ft (609.6 m)	1,039 ft³/min (29.421 m³/min)	1,113 ft³/min (31.517 m³/min)	1,326 ft³/min (37.548 m³/min)	1,426 ft³/min (40.380 m³/min)
4,000 ft (1,219.2 m)	1,174 ft³/min (33.244 m³/min)	1,323 ft³/min (37.463 m³/min)	1,486 ft³/min (42.079 m³/min)	1,648 ft³/min (46.666 m³/min)
6,000 ft (1,828.8 m)	1,310 ft³/min (37.095 m³/min)	1,573 ft³/min (44.542 m³/min)	1,646 ft³/min (46.610 m³/min)	1,946 ft³/min (55.105 m³/min)

The air velocity in the annulus must be high enough to carry the cuttings to the surface. With greater depth and increased penetration rate, an increase in the volume of air is required. An increase in volume can be attained by increased output, or pressure, from the compressors. Gas, being lighter than air, requires even higher volumes to lift the cuttings to the surface (table 13). Other variables affecting the air velocity include the amount of water entering the hole, the hole size, drill pipe diameter, and the size of the cuttings (which varies by formation).

Foam or mist drilling becomes necessary when the amount of water entering the borehole begins to fill the hole. Water is heavier than air or gas and it soon exceeds the capacity of the compressors and air circulation system to remove it. A second bad effect then occurs—cuttings stick together clogging the annulus and sticking to the drill string and bit. Stuck pipe is the frequent result.

To overcome the excess water problem, a surfactant, or foaming agent, which is a chemical similar to soap, is added to the air or gas system to initiate a mist or foam drilling operation. The crew mixes the agent and other chemicals (about 1 to 1.5 percent by volume) with water in a small tank and pumps it into the in-going air or gas stream. The foaming agent causes the water in the annulus to froth and foam. Foam is lighter than water so less air pressure is required to move the water up the hole.

Mist or foam drilling can move up to 50 bbl/h (8 m³/h) of water from the annulus. Larger amounts of water require switching to an aerated mud system or to regular water-base mud.

Aerated mud is created when both air and mud are pumped into the standpipe under controlled conditions. Air in the mud reduces hydrostatic head to the point where the rig compressors can circulate the aerated mud. An aerated mud system can also be used to overcome lost circulation problems while increasing penetration rates above those for regular drilling mud.

The mud must have low gel strength and corrosion inhibitors may be needed. Low gel strength allows the air to break out so that the mud can be recycled through the mud pumps.

Also be aware that rapid expansion of the air in the upper portion of the hole can create annular velocities as high as 2,000 or 3,000 ft/min (600 to 1,000 m/min). In most cases, casing must be set in the hole to protect formations from being eroded by such high velocities.

To summarize—

Advantages of air or gas drilling over mud

- Faster penetration rates
- Reduced bit wear
- Easy to identify type of rock being drilled
- Oil, gas, and water shows are easy to see

Disadvantages of air or gas drilling

- Cannot prevent formation fluids from entering the hole
- Large volumes of water entering hole impede drilling
- Shale can slough into hole
- Risk of explosion or fire
- Drill string corrosion

Bit Hydraulics

Hydraulics deals with the behavior of a liquid in motion. Bit hydraulics concerns the circulating pressure available at the bit to clean the bottom of the hole. The hydraulic horsepower of the circulating fluid at the bit is critical to the penetration rate because this horsepower removes the cuttings from the bottom of the hole. Hydraulic horsepower at the bit must be sufficient to efficiently remove the cuttings. If the cuttings are not removed quickly, the bit merely regrinds them instead of deepening the hole. Increasing the weight and rotary speed does not increase the rate of penetration if the hole is not cleared of cuttings.

Hydraulic horsepower is determined by pump output, which is usually measured in gallons per minute or cubic metres per minute (gpm or m³/min), and circulating pressure in psi or kPa. A change in either output or pressure directly affects hydraulic horsepower at the bit. The mud pumps generate hydraulic pressure and transmit it down the drill string, out of the bit, and up the annulus to the surface. When the mud reaches the surface, all its pressure is used up. The mud may leave the pump under thousands of pounds of pressure but at every point in the system pressure losses occur (fig. 70). Substantial pressure is lost as the mud travels through the surface equipment and down the drill string because the inside of the pipes is rough. The roughness creates friction and turbulence that reduce pressure. The largest loss, 50 to 60 percent of the total, should occur at the bit nozzles. This large pressure loss at the bit nozzles is beneficial. The mud leaves the nozzle at great velocity, creating strong turbulence at the bottom of the hole,

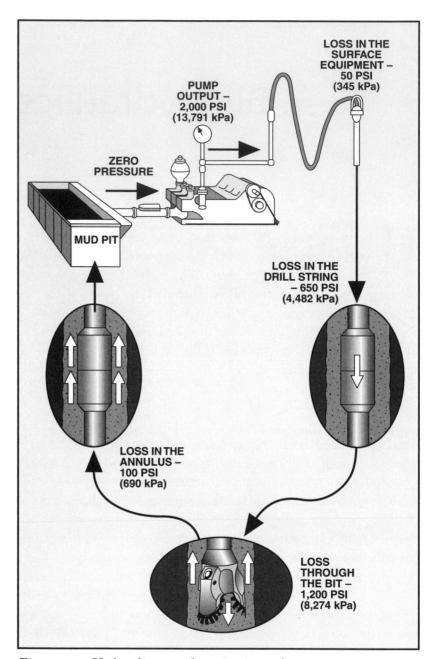

Figure 70. Hydraulic power losses in rig mud system

cleaning it and lifting the cuttings upward into the annulus where they are transported to the surface. Table 14 tabulates the pressure losses illustrated in figure 70. Keep in mind that the figures shown are the pressure losses in one circulating system in one specific drilling situation. Pressure losses vary with the system and the situation.

Table 14
Pressure Losses as Illustrated in Figure 70
with Mud Pumping Rate of 400 gal/min
at 2,000 psi (1.5 m^3/min at 13,791 kPa)

Circulation Component	Pressure Loss	Percent of Loss
Surface equipment	50 psi (345 kPa)	2.5
Drill stem	650 psi (4,482 kPa)	32.5
Bit nozzles	1,200 psi (8,274 kPa)	60.0
Return annulus	100 psi (690 kPa)	5.0
Total loss	2,000 psi (13,791 kPa)	100.0

Hydraulics Calculations

Many factors affect hydraulic pressure in a system, including pump output, depth, mud weight and viscosity, nozzle size, pipe and collar size (and whether its inside diameter is lined or unlined), and annulus volume and configuration (fig. 71). The greater the pressure loss in the system the more powerful the mud pumps must be. Drilling engineers attempt to design and maintain a circulating system that delivers 50 to 75 percent of a pump's output pressure to the bit nozzles. To achieve this goal, they must consider every variable factor that affects hydraulic pressure.

WELL NAME: 12
CUSTOMER: Y
LOCATION: Y

WELL DATA

DEPTH .... 9500 ft
HOLE SIZE .... 7.875 inches
MUD WEIGHT .... 13.50 ppg
PLAS. VISC .... 27.6 cp
YIELD PT .... 10.00 lbs/ 100 ft²

RECOMMENDED HYDRAULICS

NOZZLES   11 10 10
TFA .... 0.2462 sqin
PUMP PRESSURE .... 2998 psi
FLOW RATE .... 305 gpm
BIT HSI 6.99

HYDRAULIC DETAILS

FLOW RATE / HOLE DIAMETER .... 38.8 gpm/in.
SYSTEM PRESSURE LOSS .... 1086 psi
BIT NOZZLE PRESSURE DROP .... 1912 psi
ANNULAR PRESSURE LOSS .... 190 psi
NOZZLE PRESSURE DROP (% of available pressure) .... 63.8 %
PUMP OUTPUT HYDRAULIC POWER .... 534 hhp
BIT HYDRAULIC POWER .... 341 hhp
NOZZLE VELOCITY .... 398 ft/sec
JET IMPACT FORCE .... 849 lbs
IMPACT FORCE / HOLE AREA .... 17.43 psi
HYDROSTATIC HEAD .... 6662 psi
EQUIVALENT CIRCULATING DENSITY .... 13.88 ppg
CUTTINGS SLIP VELOCITY (chip size = .30 inch)............... 33.7 ft/min
SURFACE EQUIPMENT LENGTH .... 250 ft x 3.0 in.
AVAILABLE BUOYANT COLLAR WEIGHT .... 39304 lbs

---

Figure 71. Factors affecting rig hydraulics (Courtesy of Reed-Hycalog)

Formerly, calculations incorporating these variables required a lot of an engineer's time, but computers now allow them to quickly predict the results from changes in nozzle size, pump output, or other variables in the drilling program. Hydraulic horsepower, drilling cost per foot, and other parameters can be easily determined either in the office, utilizing a computerized drilling information network, or on site using hand-held computer-calculators.

Tables 15A and 15B show pressure losses in the drill string and can be used to determine the effect of different drill stem sizes. For example, a pump output of 320 gpm (1.21 m³/m) and 4½-in. (114.3-mm) drill pipe with 3-in. (76.20-mm) tool joint bore, pressure losses are 47 psi for every 1,000 ft (106 kPa/100 mm) of drill stem length.

The best nozzle combination can be determined from tables 16A and 16B, provided the amount of pressure available at the nozzle is known (initial pressure at pump outlet minus pressure losses before the fluid reaches the bit). Assume 1,791 psi (12,349 kPa) at the bit and pump flow rate of 320 gpm (1.21 m³/m), across from 320 (1.21) is the value 1,782 (12,287) which comes closest to, but does not exceed, 1,792 psi (12,356 kPa). The table shows that the optimum pressure will be expended through a combination of three nozzles whose inside diameter is 5⁄16 inch (7.94 mm).

The mathematical data used to derive hydraulics tables vary. Drillers and rig managers should become familiar with the methods and tables preferred by their particular company.

Table 15

Determining Pressure Loss through Drill Stem Bore

PIPE SIZE	*2⅞"			2⅞"			3½"					4"					4½"				4½"						5"	5½"				6⅝"	
TYPE OF TOOL JT.	I.F.	I.F.	S.L. H-90	S.H.	X.H.	I.F.	S.H.	I.F.	F.H. X.H.	I.F.	I.F.	F.H.	I.F.	F.H.	I.F.	I.F.	F.H.	X.H.	I.F.	I.F.	X.H.	A.S.L. F.H.	A.S.L. F.H.	D.S.L. S.H.	Acme	Reg. API	X.H.	Reg. API		Full Hole		Reg. API	F.H.
TOOL JT. BORE	2⅛"	2-7/16"	2.151"	1¾"	1⅞"	2⅛"	2⅛"	2-11/16"	2-7/16"	2-11/16"	2-11/16"	2-13/16"	3¼"	2-13/16"	3¼"	3¼"	3"	3¼"	3¾"	3¾"	3¼"	3-5/32"	3"	2-11/16"	2½"	2¼"	3¾"	2¾"	3"	3-13/16"	4"	3½"	5"
**PIPE WT.				10.40 LB.			13.30 LB.		15.50 LB.			14.00 LB.		15.70 LB.			16.60 LB.				20.00 LB.						19.50 LB.	21.90 LB.				25.20 LB.	
† LWDP WT.	6.85 LB.										9.50 LB.					11.85 LB.				13.75 LB.													
‡ TUBING	6.50 LB.										9.30 LB.					11.00 LB.				12.75 LB.													
GALLONS PER MIN.	PRESSURE LOSS—PSI PER 1,000 FEET OF DRILL STEM																																
290	323	303	304	626	596	563	191	168	228	220	111	72	67	82	77	53	39	37	34	28	46	46	48	53	53	59	21	18	16	13	13	6	4
300	344	323	325	666	635	600	204	179	243	235	119	77	71	87	82	56	42	39	37	30	49	49	51	57	57	63	23	19	17	14	13	6	5
310	365	343	345	708	675	637	217	190	258	249	126	82	76	93	87	60	44	42	39	32	52	53	54	60	60	67	24	20	18	15	14	6	5
320	388	364	366	751	716	676	230	201	274	264	134	86	80	98	92	64	47	44	41	34	55	56	58	64	64	71	25	21	19	16	15	7	5
330	410	386	387	795	758	716	243	213	290	280	142	91	85	104	97	67	50	47	44	36	58	59	61	68	68	76	27	23	20	17	16	7	6
340	434	408	409	841	802	757	257	225	306	296	150	97	90	110	103	71	53	50	46	38	61	62	65	72	72	80	28	24	21	18	17	8	6
350	458	430	432	888	846	799	272	238	323	312	158	102	95	116	109	75	56	52	49	40	65	66	68	76	76	84	30	25	22	19	18	8	6
360	483	454	455	935	892	842	286	251	341	329	167	108	100	122	114	79	59	55	51	42	68	69	72	80	80	89	32	26	24	20	19	9	6
370	508	477	479	984	938	886	301	264	358	346	175	113	105	129	120	83	62	58	54	44	72	73	76	84	84	93	33	28	25	21	20	9	7
380	534	502	503	1034	986	931	317	277	377	364	184	119	110	135	126	88	65	61	57	47	75	77	79	88	88	98	35	29	26	22	21	10	7
390	560	526	528	1085	1035	977	332	291	395	382	193	125	116	142	133	92	68	64	60	49	79	81	83	92	93	103	37	31	27	23	22	10	7
400	587	552	554	1138	1085	1024	348	305	414	400	203	131	121	149	139	96	72	67	63	51	83	85	88	97	97	108	38	32	29	24	23	11	8
410	615	578	580	1191	1136	1072	365	319	434	419	212	137	127	155	146	101	75	70	65	54	87	88	91	102	102	113	40	34	30	25	24	11	8
420	643	604	606	1246	1188	1121	381	334	454	438	222	143	133	163	152	106	78	73	68	56	91	92	96	106	106	118	42	35	31	26	25	12	9
430	671	631	634	1302	1241	1171	398	349	474	458	232	150	139	170	159	110	82	77	71	59	95	97	100	111	111	123	44	37	33	27	26	12	9
440	701	659	661	1358	1295	1223	416	364	495	478	242	156	145	177	166	115	85	80	75	61	99	101	104	116	116	129	46	39	34	29	28	13	9
450	731	687	695	1416	1350	1275	434	380	516	498	252	163	151	185	173	120	89	84	78	64	103	105	109	121	121	134	48	40	36	30	29	13	10
460	....	....	....	....	....	....	452	396	537	519	263	170	157	193	180	125	93	87	81	66	108	110	113	126	126	140	50	42	37	31	30	14	10
470	....	....	....	....	....	....	470	412	559	540	274	177	164	201	188	130	97	91	84	69	112	114	118	131	131	146	52	44	39	32	31	14	11
480	....	....	....	....	....	....	489	428	582	562	284	184	170	209	195	135	101	94	88	72	116	118	123	136	136	152	54	45	40	34	32	15	11
490	....	....	....	....	....	....	508	445	604	584	296	191	177	217	203	141	104	98	91	75	121	123	127	141	142	158	56	47	42	35	34	15	12
500	....	....	....	....	....	....	527	462	627	606	307	198	184	225	211	146	108	102	95	78	126	128	132	147	147	164	58	49	43	36	35	16	12
510	....	....	....	....	....	....	547	479	651	629	318	206	191	233	218	151	112	106	98	81	130	133	137	152	152	170	60	51	45	38	36	17	12

* API LIGHTWEIGHT DRILL PIPE AND TUBING ONLY
** REGULAR DRILL PIPE WEIGHT
† LIGHTWEIGHT DRILL PIPE WEIGHT
‡ TUBING WEIGHT

Courtesy of IADC

Table 15A
Determining Pressure Loss through Drill Stem Bore (Metric)

PRESSURE LOSS—kPa PER 100 M OF DRILL STEM

Column header block (PIPE SIZE / TYPE OF TOOL JT. / TOOL JT. BORE / weights):

PIPE SIZE	TYPE OF TOOL JT.	TOOL JT. BORE	**PIPE WT.	† LWDP WT.	‡ TUBING
*73.03 mm	I.F.	53.98 mm		10.19 kg/m	
*73.03 mm	S.L. H-90	54.64 mm			
*73.03 mm	(I.F.) 61.91 mm				9.67 kg/m
73.03 mm	S.H.	44.45 mm	15.48 kg/m		
73.03 mm	X.H.	47.63 mm	15.48 kg/m		
73.03 mm	I.F.	53.98 mm	15.48 kg/m		
88.90 mm	S.H.	53.98 mm	19.79 kg/m		
88.90 mm	I.F.	68.26 mm	23.06 kg/m	14.14 kg/m	13.84 kg/m
88.90 mm	F.H. X.H.	61.91 mm	23.06 kg/m		
88.90 mm	I.F.	68.23 mm			
101.60 mm	F.H.	71.44 mm	20.83 kg/m		
101.60 mm	I.F.	82.55 mm	23.36 kg/m		
101.60 mm	I.F.	68.26 mm		17.63 kg/m	16.37 kg/m
101.60 mm	I.F.	95.25 mm		20.46 kg/m	18.97 kg/m
114.30 mm	X.H.	82.55 mm	24.70 kg/m		
114.30 mm	F.H.	76.20 mm	24.70 kg/m		
114.30 mm	I.F.	82.55 mm			
114.30 mm	I.F.	95.25 mm			
114.30 mm	X.H.	82.55 mm	29.76 kg/m		
114.30 mm	A.S.L. F.H.	80.17 mm			
114.30 mm	A.S.L. F.H.	76.20 mm			
114.30 mm	D.S.L. S.H.	68.26 mm			
114.30 mm	Acme	63.50 mm			
114.30 mm	Reg. API	57.15 mm			
127.00 mm	X.H.	95.25 mm	29.02 kg/m		
139.70 mm	Reg. API	69.85 / 76.20 mm	35.59 kg/m		
139.70 mm	Full Hole	96.84 / 101.60 mm			
168.28 mm	Reg. API	88.90 mm	34.50 kg/m		
168.28 mm	F.H.	127.00 mm			

Pressure-loss data (kPa per 100 m of drill stem):

m³/min	*73.03 I.F.	*73.03 S.L.H-90	*73.03 tbg	73.03 S.H.	73.03 X.H.	73.03 I.F.	88.90 S.H.	88.90 I.F.	88.90 F.H.X.H.	88.90 I.F.	101.60	101.60	101.60	101.60	101.60	101.60	101.60	101.60	114.30	114.30	114.30	114.30	114.30	114.30	114.30	114.30	127.00 X.H.	139.70 R.API	139.70 F.H.	139.70	139.70	168.28 R.API	168.28 F.H.
1.10	731	686	687	1,417	1,349	1,274	432	380	516	498	251	174	152	120	88	84	77	63	104	111	104	109	120	120	134	104	48	41	36	29	29	14	9
1.14	778	731	735	1,507	1,437	1,358	461	405	550	532	269	186	161	127	95	88	84	68	111	115	109	115	129	129	143	111	52	43	38	32	29	14	11
1.17	826	776	781	1,602	1,528	1,442	491	430	584	563	285	197	172	136	100	93	88	72	118	122	120	122	136	136	152	120	54	45	41	34	32	14	11
1.21	878	824	828	1,700	1,620	1,530	520	455	620	597	303	208	181	145	106	100	93	77	124	131	127	131	145	145	161	127	57	48	43	36	34	16	11
1.25	928	873	876	1,799	1,715	1,620	550	482	656	634	321	220	192	152	113	106	100	81	131	138	134	138	154	154	172	134	61	52	45	38	36	16	14
1.29	982	923	926	1,903	1,815	1,713	582	509	692	670	339	233	204	161	120	111	104	86	138	140	147	147	163	163	181	140	63	54	48	41	38	18	14
1.33	1,036	973	978	2,010	1,914	1,808	616	539	731	706	358	247	215	170	127	118	111	91	147	154	154	156	172	172	190	149	68	57	50	43	41	18	14
1.36	1,093	1,027	1,030	2,116	2,019	1,905	647	568	772	745	378	258	226	179	134	124	115	95	154	163	163	165	181	181	201	156	72	59	54	45	43	20	14
1.40	1,150	1,079	1,084	2,227	2,123	2,005	681	597	810	783	396	272	238	188	140	131	122	95	163	165	172	172	190	190	210	163	75	63	57	48	45	20	16
1.44	1,208	1,136	1,138	2,340	2,231	2,107	717	627	853	824	416	285	249	199	147	138	129	106	170	174	179	181	199	199	222	170	79	66	59	50	48	23	16
1.48	1,267	1,190	1,195	2,455	2,342	2,211	751	660	894	864	437	301	263	208	154	145	136	111	179	183	188	190	210	210	233	179	84	70	61	52	50	23	16
1.52	1,328	1,249	1,254	2,575	2,455	2,317	788	690	937	905	459	315	274	217	163	152	143	115	188	192	199	199	220	220	244	188	86	72	66	54	52	25	18
1.55	1,392	1,308	1,314	2,695	2,571	2,426	826	722	982	948	480	330	287	229	170	158	147	122	197	206	206	210	231	231	256	197	91	77	68	57	54	25	18
1.59	1,455	1,367	1,371	2,820	2,688	2,557	862	756	1,027	991	502	344	301	240	177	165	154	127	206	210	217	217	240	240	267	206	95	79	70	59	57	27	20
1.63	1,518	1,428	1,435	2,946	2,808	2,650	901	790	1,073	1,036	525	360	315	249	186	174	161	134	215	220	226	226	251	251	278	215	100	84	75	61	59	27	20
1.67	1,591	1,491	1,496	3,073	2,931	2,768	941	824	1,120	1,082	548	376	328	260	192	181	170	138	224	229	235	235	263	263	292	224	104	88	77	66	63	29	20
1.71	1,654	1,555	1,573	3,204	3,055	2,885	982	860	1,168	1,127	570	391	342	272	201	190	177	145	233	238	247	247	274	274	303	233	109	91	81	68	66	29	23
1.74	……	……	……	……	……	……	1,023	896	1,215	1,174	595	407	355	283	210	197	183	149	244	249	256	258	285	285	317	244	113	95	84	70	68	32	23
1.78	……	……	……	……	……	……	1,064	932	1,265	1,222	620	425	371	294	220	206	190	156	253	258	267	267	296	296	330	253	118	100	88	72	70	32	25
1.82	……	……	……	……	……	……	1,107	970	1,317	1,272	643	441	385	306	229	213	199	163	263	267	278	278	308	308	344	263	122	102	91	77	72	34	25
1.86	……	……	……	……	……	……	1,150	1,007	1,367	1,322	670	459	401	319	235	222	206	170	274	278	287	287	319	319	358	274	127	106	95	79	77	34	27
1.90	……	……	……	……	……	……	1,193	1,046	1,419	1,371	695	477	416	330	244	231	215	177	285	290	299	299	333	333	371	285	131	111	97	81	79	36	27
1.93	……	……	……	……	……	……	1,238	1,084	1,473	1,423	720	493	432	342	253	240	222	183	294	301	310	310	344	344	385	294	136	115	102	86	81	38	27

* API LIGHTWEIGHT DRILL PIPE AND TUBING ONLY
** REGULAR DRILL PIPE WEIGHT
† LIGHTWEIGHT DRILL PIPE WEIGHT
‡ TUBING WEIGHT

Table 16
Determining Pressure Drop in psi Across Bit Nozzle

GAL. PER MIN.	TRI-CONE BITS														TWO-CONE BITS						
	¼"	9/32"	5/16"	11/32"	3/8"	13/32"	7/16"	15/32"	½"	9/16"	5/8"	11/16"	¾"	1"	½"	9/16"	5/8"	11/16"	¾"	13/16"	1"
100	424	265	174	.....	.....	.....	.....	.....	.....	.....	.....	.....	.....	.....	.....	.....	.....	.....	.....	.....	.....
110	514	321	210	.....	.....	.....	.....	.....	.....	.....	.....	.....	.....	.....	.....	.....	.....	.....	.....	.....	.....
120	611	382	250	171	.....	.....	.....	.....	.....	.....	.....	.....	.....	.....	.....	.....	.....	.....	.....	.....	.....
130	717	448	294	201	.....	.....	.....	.....	.....	.....	.....	.....	.....	.....	.....	.....	.....	.....	.....	.....	.....
140	832	520	341	233	.....	.....	.....	.....	.....	.....	.....	.....	.....	.....	.....	.....	.....	.....	.....	.....	.....
150	955	597	392	267	189	.....	.....	.....	.....	.....	.....	.....	.....	.....	.....	84	.....	.....	.....	.....	.....
160	1,087	679	445	304	215	.....	.....	.....	.....	.....	.....	.....	.....	.....	.....	96	.....	.....	.....	.....	.....
170	1,227	766	503	343	242	.....	.....	.....	.....	.....	.....	.....	.....	.....	.....	108	.....	.....	.....	.....	.....
180	1,375	859	563	385	272	197	.....	.....	.....	.....	.....	.....	.....	.....	194	121	.....	.....	.....	.....	.....
190	1,532	957	628	429	303	220	.....	.....	.....	.....	.....	.....	.....	.....	216	134	88	.....	.....	.....	.....
200	1,698	1,061	695	475	335	244	.....	.....	.....	.....	.....	.....	.....	.....	239	149	98	.....	.....	.....	.....
210	1,872	1,169	768	524	370	269	200	.....	.....	.....	.....	.....	.....	.....	264	164	108	.....	.....	.....	.....
220	2,054	1,283	842	575	406	295	219	.....	.....	.....	.....	.....	.....	.....	289	180	118	.....	.....	.....	.....
230	2,245	1,403	920	629	443	323	240	.....	.....	87	.....	.....	.....	.....	316	197	129	88	.....	.....	.....
240	.....	1,527	1,002	684	483	351	261	198	.....	95	.....	.....	.....	.....	344	215	141	96	.....	.....	.....
250	.....	1,657	1,088	743	524	381	283	215	.....	103	.....	.....	.....	.....	374	233	153	105	.....	.....	.....
260	.....	1,792	1,176	803	567	412	306	232	.....	112	.....	.....	.....	.....	404	252	165	113	.....	.....	.....
270	.....	1,933	1,268	866	611	445	330	250	194	120	.....	.....	.....	.....	436	272	178	122	86	.....	.....
280	.....	2,079	1,364	932	657	478	355	269	208	129	.....	.....	.....	.....	469	292	192	131	92	.....	.....
290	.....	2,230	1,463	999	705	513	381	289	223	139	92	.....	.....	.....	503	314	206	141	99	.....	.....
300	.....	2,386	1,566	1,070	754	549	408	309	239	149	98	.....	.....	.....	538	335	220	150	106	.....	.....
310	.....	.....	1,672	1,142	805	586	435	330	255	159	105	.....	.....	.....	575	358	235	161	113	.....	.....
320	.....	.....	1,782	1,217	858	625	464	352	272	169	112	.....	.....	.....	612	382	251	171	121	88	.....
330	.....	.....	1,895	1,294	913	664	493	374	289	180	119	.....	.....	.....	651	406	266	182	128	93	.....
340	.....	.....	2,011	1,374	969	705	524	397	307	191	126	.....	.....	.....	691	431	283	193	136	99	.....
350	.....	.....	2,132	1,456	1,027	747	555	421	325	203	134	91	.....	.....	733	457	300	205	144	105	.....
360	.....	.....	2,255	1,540	1,086	791	587	445	344	215	141	96	.....	.....	775	483	317	217	153	111	.....
370	.....	.....	.....	1,627	1,147	835	620	470	364	227	149	102	.....	.....	819	510	335	229	161	117	.....
380	.....	.....	.....	1,716	1,210	881	654	496	383	239	157	107	.....	.....	864	538	353	241	170	124	.....
390	.....	.....	.....	1,807	1,275	928	689	523	404	252	166	113	.....	.....	910	567	372	254	179	130	.....
400	.....	.....	.....	1,901	1,341	976	725	550	425	265	174	119	.....	.....	957	597	391	267	189	137	.....
410	.....	.....	.....	1,998	1,409	1,025	762	578	446	279	183	125	88	.....	1,005	627	411	281	198	144	.....
420	.....	.....	.....	2,096	1,482	1,076	799	606	468	292	192	131	92	.....	1,055	658	432	295	208	151	.....
430	.....	.....	.....	2,197	1,553	1,128	838	635	491	306	201	137	97	.....	1,106	689	452	309	218	158	.....
440	.....	.....	.....	.....	1,626	1,181	877	665	514	321	210	144	101	.....	1,158	722	474	324	228	166	.....
450	.....	.....	.....	.....	1,701	1,235	917	696	538	336	220	150	106	.....	1,211	755	495	338	239	174	.....
460	.....	.....	.....	.....	1,777	1,291	958	727	562	351	230	157	111	.....	1,265	789	518	354	250	181	.....
470	.....	.....	.....	.....	1,856	1,347	1,001	759	587	366	240	164	116	.....	1,321	824	540	369	261	189	.....
480	.....	.....	.....	.....	1,935	1,405	1,044	792	612	382	250	171	121	.....	1,378	859	564	385	272	197	.....
490	.....	.....	.....	.....	2,017	1,465	1,088	825	638	398	261	178	126	.....	1,436	895	587	401	283	206	90
500	.....	.....	.....	.....	2,100	1,523	1,133	859	664	414	272	186	131	.....	1,495	932	612	418	295	214	93
510	.....	.....	.....	.....	2,185	1,585	1,178	894	691	431	283	193	136	.....	1,555	970	636	435	307	223	97
520	.....	.....	.....	.....	2,271	1,647	1,225	929	718	448	294	201	142	.....	1,617	1,008	661	452	319	232	101
530	.....	.....	.....	.....	.....	1,711	1,273	965	746	465	305	209	147	.....	1,680	1,047	687	469	331	241	105
540	.....	.....	.....	.....	.....	1,777	1,321	1,002	774	483	317	217	153	.....	1,744	1,087	713	487	344	250	109
550	.....	.....	.....	.....	.....	1,843	1,370	1,039	803	501	329	225	159	.....	1,810	1,128	740	505	357	259	113
560	.....	.....	.....	.....	.....	1,911	1,421	1,078	833	520	341	233	164	.....	1,875	1,169	767	524	370	269	117

Courtesy of IADC

Table 16A
Determining Pressure Drop in kPa Across Bit Nozzle (Metric)

m³/m	TRI-CONE BITS														TWO-CONE BITS						
	6.35	7.14	7.93	8.73	9.53	10.32	11.11	11.91	12.70	14.29	15.88	17.46	19.05	25.40	12.70	14.29	15.88	17.46	19.05	20.64	25.4
0.379	2,924	1,817	1,200																		
0.417	3,544	2,213	1,448																		
0.455	4,213	2,634	1,724	1,179																	
0.493	4,944	3,089	2,027	1,448																	
0.531	5,757	3,585	2,351	1,607																	
0.569	6,585	4,116	2,703	1,841	1,303											579					
0.606	7,495	4,682	3,068	2,096	1,482											662					
0.644	8,460	5,282	3,468	2,365	1,669											745					
0.682	9,481	5,923	3,882	2,655	1,875	1,358									1,338	834					
0.720	10,563	6,599	4,330	2,958	2,089	1,517									1,489	924	607				
0.758	11,708	7,316	4,792	3,275	2,310	1,682									1,648	1,027	676				
0.796	12,907	8,060	5,295	3,613	2,551	1,855	1,379								1,820	1,131	745				
0.834	14,162	8,846	5,806	3,965	2,799	2,034	1,510								1,993	1,241	814				
0.872	15,479	9,674	6,343	4,337	3,054	2,227	1,655			600					2,179	1,358	890	607			
0.910		10,529	6,909	4,716	3,330	2,420	1,800	1,365		655					2,372	1,482	972	662			
0.948		11,425	7,502	5,123	3,613	2,627	1,951	1,482		710					2,579	1,607	1,055	724			
0.985		12,356	8,108	5,537	3,910	2,841	2,110	1,600		772					2,786	1,738	1,138	779			
1.023		13,328	8,743	5,971	4,213	3068	2,275	1,724	1,338	827					3,006	1,875	1,227	841	593		
1.061		14,335	9,405	6,426	4,530	3,296	2,448	1,855	1,434	890					3,234	2,013	1,324	903	634		
1.100		15,376	10,087	6,888	4,861	3,537	2,627	1,993	1,538	958	634				3,468	2,165	1,420	972	683		
1.137		16,452	10,798	7,378	5,199	3,785	2,813	2,131	1,648	1,027	676				3,710	2,310	1,517	1,034	731		
1.175			11,528	7,874	5,551	4,041	2,999	2,775	1,758	1,096	724				3,965	2,468	1,620	1,110	779		
1.213			12,287	8,391	5,916	4,309	3,199	2,427	1,875	1,165	772				4,220	2,634	1,731	1,179	834	607	
1.251			13,066	8,922	6,295	4,578	3,399	2,579	1,993	1,241	82				4,482	2,799	1,834	1,255	883	641	
1.289			13,866	9,474	6,681	4,861	3,613	2,737	2,117	1,317	869				4,764	2,972	1,944	1,331	938	683	
1.327			14,700	10,039	7,081	5,151	3,827	2,903	2,241	1,400	924	628			5,054	3,151	2,069	1,414	993	724	
1.364			15,548	10,618	7,488	5,454	4,047	3,068	2,372	1,482	972	662			5,344	3,330	2,186	1,496	1,055	765	
1.402				11,218	7,909	5,757	4,275	3,241	2,510	1,565	1,027	703			5,647	3,517	2,310	1,579	1,110	807	
1.440				11,832	8,343	6,075	4,509	3,420	2,641	1,648	1,083	738			5,957	3,710	2,434	1,662	1,172	855	
1.478				12,459	8,791	6,399	4,751	3,606	2,786	1,738	1,145	779			6,275	3,910	2,565	1,751	1,234	896	
1.516				13,107	9,246	6,730	4,999	3,792	2,930	1,827	1,200	821			6,599	4,116	2,696	1,841	1,303	945	
1.554				13,776	9,715	7,067	5,254	3,985	3,075	1,924	1,262	862	607		6,930	4,323	2,834	1,938	1,365	993	
1.592				14,452	10,218	7,419	5,509	4,178	3,227	2,013	1,324	903	634		7,274	4,537	2,979	2,034	1,434	1,041	
1.630				15,148	10,708	7,778	5,778	4,378	3,386	2,110	1,386	945	669		7,626	4,751	3,117	2,131	1,503	1,089	
1.668					11,211	8,143	6,047	4,585	3,544	2,213	1,448	993	696		7,984	4,978	3,268	2,234	1,572	1,145	
1.706					11,728	8,515	6,323	4,799	3,710	2,317	1,517	1,034	731		8,350	5,206	3,413	2,331	1,648	1,200	
1.743					12,252	8,902	6,605	5,013	3,875	2,420	1,586	1,083	766		8,722	5,440	3,572	2,441	1,724	1,248	
1.781					12,797	9,288	6,902	5,233	4,047	2,524	1,655	1,131	800		9,108	5,682	3,723	2,544	1,800	1,303	
1.819					13,342	9,688	7,198	5,461	4,220	2,634	1,724	1,179	834		9,501	5,923	3,889	2,655	1,875	1,358	
1.857					13,907	10,101	7,502	5,688	4,385	2,744	1,800	1,227	869		9,901	6,171	4,047	2,765	1,951	1,420	621
1.895					14,480	10,501	7,812	5,923	4,578	2,855	1,875	1,283	903		10,308	6,426	4,220	2,882	2,034	1,476	641
1.933					15,066	10,929	8,122	6,164	4,765	2,972	1,951	1,331	938		10,722	6,688	4,385	2,999	2,117	1,538	669
1.971					15,659	11,356	8,446	6,406	4,951	3,089	2,027	1,386	979		11,149	6,950	4,558	3,117	2,200	1,600	696
2.009						11,797	8,777	6,654	5,144	3,206	2,103	1,441	1,014		11,584	7,219	4,737	3,234	2,282	1,662	724
2.047						12,252	9,108	6,909	5,337	3,330	2,186	1,496	1,055		12,025	7,495	4,916	3,358	2,372	1,724	752
2.085						12,708	9,446	7,164	5,537	3,454	2,269	1,551	1,096		12,480	7,778	5,102	3,482	2,462	1,786	779
2.122						13,176	9,798	7,433	5,744	3,585	2,351	1,607	1,131		12,928	8,060	5,289	3,613	2,551	1,855	807

To summarize—

- Hydraulic horsepower affects how well bottom of hole is cleaned; if hole is not cleaned of cuttings, ROP decreases.
- Pump output in gpm or m³/min and circulating pressure in psi or kPa determines hydraulic horsepower.
- Pressure losses in drill string should be greatest at bit nozzles because mud leaves the bit with great speed (velocity) to clean hole.

Formation Properties

The nature of the formation being drilled greatly influences the drilling rate and other important aspects of the overall drilling operation. Petroleum geologists gather information from nearby wells and predict formation depths (tops), rock character, and geological hazards. A geologist prepares a well prognosis setting out this information and the rig manager or drilling superintendent will usually have access to it. The bit program may be based in part on the information in the prognosis. A geologist will often be on site to examine well cuttings, call formation tops, help pick casing, logging, and coring points, as well as the total depth (TD). The operator may also engage a mud logging company to provide full time, continuous monitoring of the mud stream for shows of oil and gas and to identify the formations being drilled.

In general, porous and permeable formations drill faster than impermeable formations. If the bit is drilling an impermeable zone and it enters a permeable zone, a *drilling break* may occur. A drilling break is an increase in the rate of penetration. If a drilling break occurs in a possible pay horizon, drilling may be stopped and the cuttings, along with any shows of hydrocarbons in the mud, circulated to the surface. Formation testing may be conducted at that point or an evaluation may be deferred until logs are run.

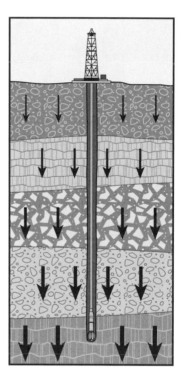

Figure 72. Overburden (formation pressure increases with depth)

Overburden pressure increases with depth and a corresponding increase in a rock's compressive strength with depth occurs (fig. 72). Because rock strength increases with depth, the driller must increase the weight on the bit; otherwise, the bit cannot overcome the compressive strength of the rock.

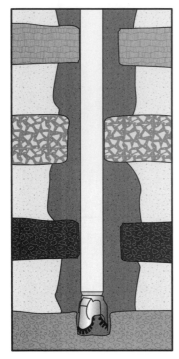

Figure 73. Drilling through alternating hard and soft zones can produce offset ledges.

The ability to drill different types of rock varies greatly. This variation occurs because rocks have different compressive strengths. Chalk and soft limestones have compressive strengths of 5,000 psi (34,475 kPa), or less, while granite and quartzite have compressive strengths of 45,000 psi (310,275 kPa), or more. Generally, older rocks in the Mid-Continent or Appalachian areas of the U.S. drill slower than younger Gulf Coast rocks, for instance.

The reaction of some formations to water-base mud can slow drilling. Shale and clay is especially prone to sloughing, swelling, or forming a sticky mixture. In many cases, the circulating drilling mud has difficulty in cleaning, or removing, this sticky and sloughing shale or clay from the bottom of the hole. A balled-up bit or stuck pipe is common in such problem formations. A switch to expensive oil-base mud may become necessary. Again, the geological prognosis should warn the contractor of the risks associated with certain formations.

The effects of rock hardness, softness, or abrasiveness on the rate of penetration greatly influences bit selection. Bit design features such as cone offset, bit type (tooth, insert, or PDC), hardfacing, bearings, etc., must be carefully chosen in order to keep trip and bit costs to a minimum.

Alternating layers of soft and hard rock can cause the bit to be deflected from its normal course. It tends to drift and follow the easiest path, causing changes in the angle of the hole that create offset ledges (fig. 73). Gauge rounding and shirttail damage are indicators of offset ledges that require extra care going in or out of the hole.

Formation dip, a formation's inclination from horizontal, can cause the hole to deviate. Areas where dipping formations make it difficult to drill a vertical hole are often referred to as "crooked hole country." Hole deviation in dipping formations can generally be controlled by use of heavier, stiffer drill collars, stabilizers, and less weight on the bit. Modern measurement-while-drilling (MWD) technology allows much closer monitoring of hole deviation than previously possible.

Where formation dip is less than 45 degrees, the bit will tend to walk up-dip because the up-dip edge of the bit will bear more weight and thus drill faster (fig. 74). The rig location can be planned to allow the bit to drift steadily up-dip and still hit the target (fig. 75).

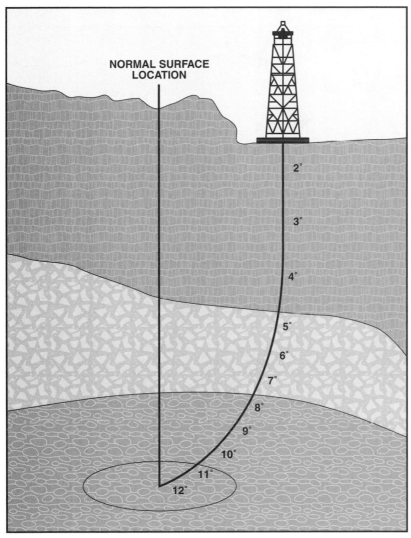

Figure 75. *Faster drilling may be achieved if the bit is allowed to drift into the target area.*

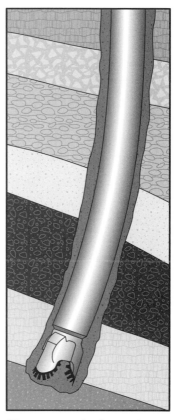

Figure 74. In formations with dip of less than 45 degrees, the bit tends to drift or walk up-dip.

In formations where the dip is 45 degrees or more, the bit will tend to slide (deviate) down the bedding planes because the rock will fracture more easily along the bed boundaries (fig. 76). Deviation in dipping beds is a bigger problem in formations with laminar bedding of alternating hard and soft ledges than it is in thick, uniform, nonbedded formations.

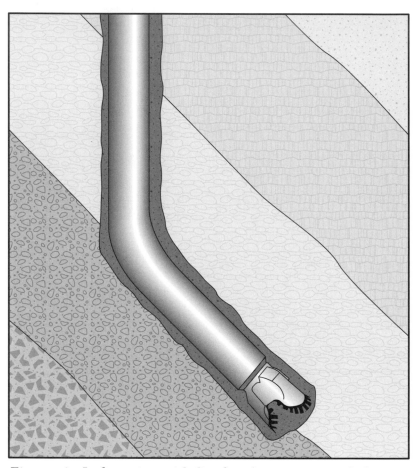

Figure 76. In formations with dip of 45 degrees or more, the bit tends to drift (slide) down-dip along bedding planes.

To summarize—

- Nature of formation determines ROP and other aspects of drilling.
- Generally, porous and permeable formations drill faster than nonporous, impermeable formations.
- Rock strength increases with depth, so WOB must be increased to overcome strength of deep rocks.
- Generally, older rocks drill slower than younger rocks.
- A formation's reaction to water-base mud can slow drilling rate.
- Formation dip can cause crooked holes to be drilled.

New Technology

Ongoing research conducted by companies, trade organizations, and government will impact future drilling operations. Some current research projects show promise and are likely to be adopted by the industry within 3 to 5 years. While it is not possible to predict the results of research efforts, or of some future scientific breakthrough, certain existing trends can be projected into the future. For instance, average well depth will increase and large fluctuations in rig employment will occur. Contractors must therefore remain adaptable and willing to train their crews to use new technology as it evolves.

Computers will increasingly influence all industries, including the drilling industry. The computer is now a basic tool in the offshore, in deep drilling, in directional drilling, and in large, multiwell operations. Integrated computer systems will probably be in place on all rigs drilling below 10,000 ft, or drilling directional or horizontal holes, in the near future. These sophisticated information systems will allow drilling engineers and management to control rig operations from the office, no matter how remote. Rigs drilling shallow holes on land and involved in intermittent projects will likely continue to drill without too much computerized control because of the cost involved.

Well planning will be improved by increased use of computers. With a computer it is possible to record and review the drilling records of thousands of wells in a large basin-size area. Drilling operations can be grouped into related functions to determine all the best procedures. The best procedures can then be incorporated into an optimized drilling program.

As improvements are made to downhole motors, turbodrills, and top drives, their use will increase, possibly replacing standard rotary drilling offshore and on larger land rigs. Also, the use of coiled tubing with downhole motors may prove to be a practical drilling procedure.

Downhole electronics, such as MWD techniques, will continue to evolve, providing more useful data while drilling proceeds. Downhole conditions, including rock character, as well as all drilling variables that affect penetration rate, will be routinely monitored as the bit is turning. Logging while drilling (LWD) will undoubtedly improve in capability and increase in use. Directional and horizontal drilling activity will increase because these improvements in downhole electronics will ultimately lead to reduced costs.

Underbalanced drilling shows promise for improving penetration rates in certain situations. This approach to drilling depends on maintaining well control while keeping the hydrostatic pressure at or below formation pressure. Penetration rates improve under such conditions because drilling rates increase as mud weights decrease (see fig. 62). Several industry groups are actively involved in drilling projects using underbalanced mud systems and the practice is likely to increase.

Drill bit technology will progress in several ways. Developments in metallurgy, PDC and TSP bit design, bit bearings, and automatic drilling technology will extend bit life. Research in disk-cutter bits holds promise. Disk-cutter bits use technology employed in large tunnel-boring bits—technology that may be adapted to smaller rotary drilling bits. Early field trials using single cone bits for drilling small holes are encouraging. The single cone is much larger, with heavier cutters and bearings, than the cones in similar size tricone bits (fig. 77). Given time for additional development, the single cone bit may become common in slim-hole applications. Depending on long-term research results, laser or helical bits could someday replace current designs.

Improvements in drilling fluids, additives, and lost circulation control, will lead to more drilling in problem areas where potential has been overlooked in the past. In the future, the well-trained crew, using the latest developments in drilling technology, will routinely handle the deep, corrosive, deviation, or high temperature drilling problems of today.

*Figure 77.   Single-cone bit for drilling small diameter holes
(Courtesy of RBI-Gearhart)*

To summarize—
- Computers will play an increasing role in drilling.
- Use of coiled tubing drilling will likely increase.
- MWD techniques will continue to improve.
- Use of underbalanced drilling to increase ROP will likely increase.
- Improvements in bit technology will increase bit life.
- Improvements in drilling fluid technology will allow problem areas to be drilled that could not be drilled in the past.

Glossary

abrasion *n*: wearing away by friction.

absolute permeability *n*: a measure of a single fluid (such as water, gas, or oil) to flow through a rock formation when the rock is totally filled (saturated) with the single fluid. The permeability measure of a rock filled with a single fluid is different from the permeability measure of the same rock filled with two or more fluids. Compare *effective permeability*.

absolute porosity *n*: the percentage of the total bulk volume of a rock sample that is composed of pore spaces or voids. See *porosity*.

air drilling *n*: a method of rotary drilling that uses compressed air as the circulating medium. The conventional method of removing cuttings from the wellbore is to use a flow of water or drilling mud. Compressed air removes the cuttings with equal or greater efficiency. The rate of penetration is usually increased considerably when air drilling is used. However, a principal problem in air drilling is the penetration of formation containing water, since the entry of water into the system reduces the ability of the air to remove the cuttings.

annular space *n*: the space between two concentric circles. In the petroleum industry, it is usually the space surrounding a pipe in the wellbore, or the space between tubing and casing, or the space between tubing and the wellbore; sometimes termed the annulus.

annulus *n*: see *annular space*.

antiwhirl bit *n*: a drill bit, usually a polycrystalline diamond bit, that is designed to prevent the bit's drilling a spiral-shaped hole because it whirls off-center as it rotates. See *bit whirl*.

API gravity *n*: the measure of the density or gravity of liquid petroleum products in the United States; derived from relative density in accordance with the following equation:

API gravity at 60°F = [141.5 ÷ relative density 60/60°F] − 131.5

API gravity is expressed in degrees, 10°API being equivalent to 1.0, the specific gravity of water. See *gravity*.

ball up *v*: to collect a mass of sticky consolidated material, usually drill cuttings, on drill pipe, drill collars, bits, and so forth. A bit with such material attached to it is called a balled-up bit. Balling up is frequently the result of inadequate pump pressure or insufficient drilling fluid.

barite *n*: barium sulfate, a mineral frequently used to increase the weight or density of drilling mud. Its specific gravity is 4.2 (i.e., it is 4.2 times denser than water). See *barium sulfate, mud.*

barium sulfate *n*: a chemical compound of barium, sulfur, and oxygen ($BaSO_4$), which may form a tenacious scale that is very difficult to remove. Also called barite.

barrel (bbl) *n*: a measure of volume for petroleum products in the United States. One barrel is the equivalent of 42 U.S. gallons or 0.15899 cubic metres (9,702 cubic inches). One cubic metre equals 6.2897 barrels.

baryte *n*: variation of barite. See *barite.*

bearing *n*: 1. an object, surface, or point that supports. 2. a machine part in which another part (such as a journal or pin) turns or slides.

bent housing *n*: a special housing for the positive-displacement downhole mud motor, which is manufactured with a bend of 1° to 3° to facilitate directional drilling.

bentonite *n*: a colloidal clay, composed primarily of montmorillonite, that swells when wet. Because of its gel-forming properties, bentonite is a major component of water-base drilling muds. See *gel, mud.*

bent sub *n*: a short, cylindrical device installed in the drill stem between the bottommost drill collar and a downhole motor. Its purpose is to deflect the downhole motor off vertical to drill a directional hole. See *drill stem.*

bhhp *abbr*: bit hydraulic horsepower.

BHP *n*: bottomhole pressure.

BHT *n*: bottomhole temperature.

bit *n*: the cutting or boring element used in drilling oil and gas wells. The bit consists of a cutting element and a circulating element. The cutting element is steel teeth, tungsten carbide buttons, industrial diamonds, or polycrystalline diamond compacts (PDCs). The circulating element permits the passage of drilling fluid and utilizes the hydraulic force of the fluid stream to improve drilling rates. In rotary drilling, several drill collars are joined to the bottom end of the drill pipe column, and the bit is attached to the end of the drill collars.

bit cone *n*: on a roller cone bit, a cone-shaped steel device from which the manufacturer either mills or forges steel teeth, or into which the manufacturer inserts tungsten carbide buttons. Most roller cone bits have three cones, which roll, or rotate, on bearings as the bit rotates. As the cones roll over the formation, the cutters on the cone scrape or gouge the formation to remove the rock.

bit cutter *n*: the cutting elements of a bit.

bit hydraulic horsepower *n*: the measure of hydraulic power expended through the bit nozzles for cleaning the bit cutters and the hole bottom.

bit program *n*: a plan for the expected number and types of bits that are to be used in the drilling of a well. The bit program takes into account all the factors that affect bit performance so that reliable cost calculations can be made.

bit record *n*: a report that lists each bit used during a drilling operation, giving the type, the footage it drilled, the formation it penetrated, its condition, and so on.

bit whirl *n*: the motion a bit makes when it does not rotate around its center but instead drills with a spiral motion. It usually occurs to a bit drilling in a soft or medium soft formation when the driller does not apply enough weight or does not rotate the bit fast enough. A whirling bit drills an overgauge hole (a hole larger than the diameter of the bit) and causes the bit to wear abnormally.

block *n*: any assembly of pulleys on a common framework; in mechanics, one or more pulleys, or sheaves, mounted to rotate on a common axis. The crown block is an assembly of sheaves mounted on beams at the top of the derrick or mast. The drilling line is reeved over the sheaves of the crown block alternately with the sheaves of the traveling block, which is hoisted and lowered in the derrick or mast by the drilling line. When elevators are attached to a hook on a conventional traveling block, and when drill pipe is latched in the elevators, the pipe can be raised or lowered in the derrick or mast. See *crown block, traveling block*.

blooey line *n*: the discharge pipe from a well being drilled by air drilling. The blooey line is used to conduct the air or gas used for circulation away from the rig to reduce the fire hazard as well as to transport the cuttings a suitable distance from the well. See *air drilling*.

blowout *n*: an uncontrolled flow of gas, oil, or other well fluids into the atmosphere or into an underground formation. A blowout, or gusher, can occur when formation pressure exceeds the pressure applied to it by the column of drilling fluid. See *kick*.

borehole *n*: a hole made by drilling or boring; a wellbore.

bottomhole assembly *n*: the portion of the drilling assembly below the drill pipe. It can be very simple—composed of only the bit and drill collars—or it can be very complex and made up of several drilling tools.

bottomhole pressure *n*: the pressure at the bottom of a borehole. It is caused by the hydrostatic pressure of the wellbore fluid and, sometimes, by any back-pressure held at the surface, as when the well is shut in with blowout preventers. When mud is being circulated, bottomhole pressure is the hydrostatic pressure plus the remaining circulating pressure required to move the mud up the annulus.

bradding *n*: a condition in which the weight on a bit tooth has been so great that the tooth has dulled until its softer inner portion caves over the harder case area.

button bit *n*: a drilling bit with tungsten carbide inserts on the cones that resemble plugs or buttons.

C

casing *n*: steel pipe placed in an oil or gas well to prevent the wall of the hole from caving in, to prevent movement of fluids from one formation to another, and to improve the efficiency of extracting petroleum if the well is productive. A joint of casing may be 16 to 48 feet (4.8 to 14.6 metres) long and from 4.5 to 20 inches (11.4 to 50.8 centimetres) in diameter. Casing is made of many types of steel alloy, which vary in strength, corrosion resistance, and so on.

caustic soda *n*: sodium hydroxide, NaOH. It is used to maintain an alkaline pH in drilling mud and in petroleum fractions.

caving *n*: collapsing of the walls of the wellbore. Also called sloughing.

cement *n*: a powder consisting of alumina, silica, lime, and other substances that hardens when mixed with water. Extensively used in the oil industry to bond casing to the walls of the wellbore.

chip hold-down effect *n*: the holding of formation rock chips in place as a result of high differential pressure in the wellbore (i.e., pressure in the wellbore is greater than pressure in the formation). This effect limits the cutting action of the bit by retarding circulation of bit cuttings out of the hole.

circulating pressure *n*: the pressure generated by the mud pumps and exerted on the drill stem.

circulation *n*: the movement of drilling fluid out of the mud pits, down the drill stem, up the annulus, and back to the mud pits. See *normal circulation, reverse circulation*.

clay *n*: 1. a term used for particles smaller than 1/256 millimetre (4 microns) in size, regardless of mineral composition. 2. a group of hydrous aluminum silicate minerals (clay minerals). 3. a sediment of fine clastics.

clear water drilling *n*: drilling operations in which plain water (usually salt water) is used as the circulating fluid.

closed system *n*: in circulation, a system in which no drilling fluid is discarded to a reserve pit. Drilling companies may use closed systems when environmental regulations do not permit any contaminants to be released. Companies are also discovering that closed systems can be economical when using expensive drilling muds.

compression *n*: the act or process of squeezing a given volume of gas into a smaller space.

compressive strength *n*: the degree of resistance of a material to a force acting along one of its axes in a manner tending to collapse it; expressed in pounds of force per square inch (psi) of surface affected or in kilopascals of force of surface affected.

compressor *n*: a device that raises the pressure of a compressible fluid such as air or gas. Compressors create a pressure differential to move or compress a vapor or a gas, consuming power in the process. They may be positive-displacement compressors or nonpositive-displacement compressors.

cone offset *n*: the amount by which lines drawn through the center of each cone of a bit fail to meet in the center of the bit. For example, in a roller cone bit with three cones, three lines can be drawn through the center of each cone and extended to the center of the bit. If these cone centerlines do not meet in the bit's center, the cones are said to be offset. In general, bits designed for drilling soft formations have more offset than cones for hard formations, because offset affects the angle at which the bit teeth contact the formation. Since soft formations require a gouging and scraping action by bit teeth, high offset achieves the necessary action.

cone shell *n*: that part of the cone of a roller cone bit out of which the teeth are milled or into which tungsten carbide inserts are placed and inside of which the bearings are housed.

cone skidding *n*: locking of a cone on a roller cone bit so that it will not turn when the bit is rotating. Cone skidding results in the flattening of the surface of the cone in contact with the bottom of the hole.

control well *n*: a well previously drilled in an area of drilling interest, the data from which may be a reliable source of information in the planning of a new well.

corrosion *n*: any of a variety of complex chemical or electrochemical processes, e.g., rust, by which metal is destroyed through reaction with its environment.

corrosion-control agent *n*: a chemical added to drilling fluid to minimize corrosion to the drill stem.

crown block *n*: an assembly of sheaves mounted on beams at the top of the derrick or mast and over which the drilling line is reeved.

cutters *n pl*: 1. on a bit used on a rotary rig, the elements on the end (and sometimes the sides) of the bit that scrapes, gouges, or otherwise removes the formation to make hole. 2. the parts of a reamer that actually contact the wall of the hole and open it to full gauge. A three-point reamer has three cutters; a six-point reamer has six cutters. Cutters are available for different formations.

cuttings *n pl*: the fragments of rock dislodged by the bit and brought to the surface in the drilling mud. Washed and dried cuttings samples are analyzed by geologists to obtain information about the formations drilled.

D

daily drilling report *n*: a record made each day of the operations on a working drilling rig and, traditionally, phoned or radioed in to the office of the drilling company every morning. Also called morning report.

day rate *n*: an hourly or daily contract price the operator agrees to pay for use of rig, crew, and specified equipment. A day rate contract allows the operator to directly supervise the daily drilling operations.

degasser *n*: the device used to remove unwanted gas from a liquid, especially from drilling fluid.

demulsifier *n*: a chemical with properties that cause the water droplets in a water-in-oil emulsion to merge and settle out of the oil, or oil droplets in an oil-in-water emulsion to coalesce, when the chemical is added to the emulsion. Also called emulsion breaker.

density *n*: the mass or weight of a substance per unit volume. For instance, the density of a drilling mud may be 10 pounds per gallon, 74.8 pounds per cubic foot, or 1,198.2 kilograms per cubic metre. Specific gravity, relative density, and API gravity are other units of density.

derrickhand *n*: the crew member who handles the upper end of the drill string as it is being hoisted out of or lowered into the hole. He or she is also responsible for the circulating machinery and the conditioning of the drilling fluid.

desander *n*: a centrifugal device for removing sand from drilling fluid to prevent abrasion of the pumps. It may be operated mechanically or by a fast-moving stream of fluid inside a special cone-shaped vessel, in which case it is sometimes called a hydrocyclone. Compare *desilter*.

desilter *n*: a centrifugal device for removing very fine particles, or silt, from drilling fluid to keep the amount of solids in the fluid at the lowest possible point. Usually, the lower the solids content of mud, the faster is the rate of penetration. The desilter works on the same principle as a desander. Compare *desander*.

diameter *n*: the distance across a circle, measured through its center. In the measurement of pipe diameters, the inside diameter is that of the interior circle and the outside diameter that of the exterior circle.

diamond bit *n*: a drill bit that has small industrial diamonds embedded in its cutting surface. Cutting is performed by the rotation of the very hard diamonds over the rock surface.

differential pressure *n*: the difference between two fluid pressures; for example, the difference between the pressure in a reservoir and in a wellbore drilled in the reservoir, or between atmospheric pressure at sea level and at 10,000 feet (3,048 metres). Also called pressure differential.

differential sticking *n*: a condition in which the drill stem becomes stuck against the wall of the wellbore because part of the drill stem (usually the drill collars) has become embedded in the filter cake. Necessary conditions for differential-pressure sticking, or wall sticking, are a permeable formation and a pressure differential across a nearly impermeable filter cake and drill stem. Also called wall sticking. See *differential pressure, filter cake.*

directional drilling *n*: intentional deviation of a wellbore from the vertical. Although wellbores are normally drilled vertically, it is sometimes necessary or advantageous to drill at an angle from the vertical. Controlled directional drilling makes it possible to reach subsurface areas laterally remote from the point where the bit enters the earth. It often involves the use of deflection tools.

discharge line *n*: a line through which drilling mud travels from the mud pump to the standpipe on its way to the wellbore.

dogleg *n*: an abrupt change in direction in the wellbore, frequently resulting in the formation of a keyseat.

downhole *adj, adv*: pertaining to the wellbore.

downhole motor *n*: a drilling tool made up in the drill string directly above the bit. It causes the bit to turn while the drill string remains fixed. It is used most often as a deflection tool in directional drilling, where it is made up between the bit and a bent sub (or, sometimes, the housing of the motor itself is bent). Two principal types of downhole motor are the positive-displacement motor and the downhole turbine motor. Also called mud motor.

downtime *n*: 1. time during which rig operations are temporarily suspended because of repairs or maintenance. 2. time during which a well is off production.

drag bit *n*: any of a variety of drilling bits that have no moving parts. As they are rotated on bottom, elements of the bit make hole by being pressed into the formation and being dragged across it.

drill collar *n*: a heavy, thick-walled tube, usually steel, used between the drill pipe and the bit in the drill stem to provide weight on the bit.

driller *n*: the employee directly in charge of a drilling or workover rig and crew. The driller's main duty is operation of the drilling and hoisting equipment, but this crew member is also responsible for downhole condition of the well, operation of downhole tools, and pipe measurements.

drill-in fluid *n*: a drilling fluid specially formulated to minimize formation damage as the borehole penetrates the producing zone. See *formation damage.*

drilling contractor *n*: an individual or group of individuals who own a drilling rig and contract their services for drilling wells to a certain depth.

drilling crew *n*: a driller, a derrickhand, and two or more helpers who operate a drilling or workover rig for one tour each day.

drilling fluid *n*: a liquid, air, or natural gas that is circulated through the wellbore during rotary drilling operations.

drilling mud *n*: specially compounded liquid circulated through the wellbore during rotary drilling operations. See *drilling fluid, mud*.

drilling rate *n*: the speed with which the bit drills the formation; usually called the rate of penetration (ROP).

drill-off test *n*: a method of determining optimum weight on bit and overall bit performance. A given amount of weight is put on the bit and the drawworks brake is tied off so that no more weight is applied to the bit as it drills. The time it takes for the bit to stop drilling ahead with the given amount of weight is measured. Different weights are applied and the times are compared to determine which amount is best.

drill pipe *n*: seamless steel or aluminum pipe made up in the drill stem between the kelly or top drive on the surface and the drill collars on the bottom. During drilling, it is usually rotated while drilling fluid is circulated through it. Joints of pipe about 30 feet long are coupled together by means of tool joints.

drill stem *n*: all members in the assembly used for rotary drilling from the swivel to the bit, including the kelly, the drill pipe and tool joints, the drill collars, the stabilizers, and various specialty items. Compare *drill string*.

drill string *n*: the column, or string, of drill pipe with attached tool joints that transmits fluid and rotational power from the kelly to the drill collars and the bit. Often, especially in the oil patch, the term is loosely applied to include both drill pipe and drill collars. Compare *drill stem*.

E

effective permeability *n*: a measure of the ability of a single fluid to flow through a rock when another fluid is also present in the pore spaces. Compare *absolute permeability, relative permeability*.

effective porosity *n*: the percentage of the bulk volume of a rock sample that is composed of interconnected pore spaces that allow the passage of fluids through the sample. See *porosity*.

efficiency *n*: the ratio of useful energy produced by an engine to the energy put into it.

electric well log *n*: a record of certain electrical characteristics (such as resistivity and conductivity) of formations traversed by the borehole. It is made to identify the formations, determine the nature and amount of fluids they contain, and estimate their depth. Also called an electric log or electric survey.

emulsifier *n*: a material that causes water and oil to form an emulsion. Water normally occurs separately from oil; if, however, an emulsifying agent is present, the water becomes dispersed in the oil as tiny droplets. Or, rarely, the oil may be dispersed in the water. In either case, the emulsion must be treated to separate the water and the oil.

emulsion *n*: a mixture in which one liquid, termed the dispersed phase, is uniformly distributed (usually as minute globules) in another liquid, called the continuous phase or dispersion medium. In an oil-in-water emulsion, the oil is the dispersed phase and the water the dispersion medium; in a water-in-oil emulsion, the reverse holds.

explosion-proof motor *n*: a motor with an enclosure designed to contain an internal explosion and to prevent ignition of surrounding gases or vapors by sparks that may occur in the motor.

F

fatigue *n*: the tendency of material such as a metal to break under repeated cyclic loading at a stress considerably less than the tensile strength shown in a static test.

filter *n*: a porous medium through which a fluid is passed to separate particles of suspended solids from it.

filter cake *n*: 1. compacted solid or semisolid material remaining on a filter after pressure filtration of mud with a standard filter press. Thickness of the cake is reported in thirty-seconds of an inch or in millimetres. 2. the layer of concentrated solids from the drilling mud or cement slurry that forms on the walls of the borehole opposite permeable formations; also called wall cake or mud cake.

filtration loss *n*: the escape of the liquid part of a drilling mud into permeable formations.

fishtail bit *n*: a drilling bit with cutting edges of hard alloys. Developed about 1900, and first used with the rotary system of drilling, it is still useful in drilling very soft formations. Also called a drag bit.

flare *v*: to dispose of surplus combustible vapors by burning them in the atmosphere.

flocculation *n*: the coagulation of solids in a drilling fluid, produced by special additives or by contaminants.

fluid loss *n*: the unwanted migration of the liquid part of the drilling mud or cement slurry into a formation, often minimized or prevented by the blending of additives with the mud or cement.

foam *n*: a two-phase system, similar to an emulsion, in which the dispersed phase is a gas or air.

foaming agent *n*: a chemical used to lighten the water column in gas wells, in oilwells producing gas, and in drilling wells in which air or gas is used as the drilling fluid so that the water can be forced out with the air or gas to prevent its impeding the production or drilling rate. See *mist drilling*.

footage rate *n*: a footage rate contract specifies a price per foot the operator will pay the contractor for drilling a hole to an agreed depth. The size, deviation limits, and other specifications are set out in the contract. The contractor normally supervises drilling operations except when the operator temporarily takes over operations on a day rate basis.

formation *n*: a bed or deposit composed throughout of substantially the same kind of rock; often a lithologic unit. Each formation is given a name, frequently as a result of the study of the formation outcrop at the surface and sometimes based on fossils found in the formation.

formation damage *n*: the reduction of permeability in a reservoir rock caused by the invasion of drilling fluid and treating fluids to the section adjacent to the wellbore. It is often called skin damage.

formation fluid *n*: fluid (such as gas, oil, or water) that exists in a subsurface rock formation.

formation pressure *n*: the force exerted by fluids in a formation, recorded in the hole at the level of the formation with the well shut in. Also called reservoir pressure or shut-in bottomhole pressure.

friction *n*: resistance to movement created when two surfaces are in contact. When friction is present, movement between the surfaces produces heat.

friction loss *n*: a reduction in the pressure of a fluid caused by its motion against an enclosed surface (such as a pipe). As the fluid moves through the pipe, friction between the fluid and the pipe wall and within the fluid itself creates a pressure loss. The faster the fluid moves, the greater are the losses.

galling *adj*: the result of the sticking or adhesion of two mating surfaces of metal, not protected by a film of lubricant, and tearing due to lateral displacement.

G

gas *n*: a compressible fluid that completely fills any container in which it is confined. Technically, a gas will not condense when it is compressed and cooled, because a gas can exist only above the critical temperature for its particular composition. Below the critical temperature, this form of matter is known as a vapor, because liquid can exist and condensation can occur. Sometimes the terms "gas" and "vapor" are used interchangeably. The latter, however, should be used for those streams in which condensation can occur and that originate from, or are in equilibrium with, a liquid phase.

gauge *n*: 1. the diameter of a bit or the hole drilled by the bit. 2. a device (such as a pressure gauge) used to measure some physical property. *v*: to measure size, volume, depth, or other measurable property.

gauge cutters *n pl*: the teeth or tungsten carbide inserts in the outermost row on the cones of a bit, so called because they cut the outside edge of the hole and determine the hole's gauge or size. Also called heel teeth.

gel *n*: a semisolid, jellylike state assumed by some colloidal dispersions at rest. When agitated, the gel converts to a fluid state. Also a nickname for bentonite. *v*: to take the form of a gel; to set.

gel strength *n*: a measure of the ability of a colloidal dispersion to develop and retain a gel form, based on its resistance to shear. The gel, or shear, strength of a drilling mud determines its ability to hold solids in suspension. Sometimes bentonite and other colloidal clays are added to drilling fluid to increase its gel strength.

geological correlation *n*: the relating of subsurface information obtained from one well to that of others.

Geolograph™ *n*: trade name for a patented device that automatically records the rate of penetration and depth during drilling.

gpm *abbr*: 1. gallons per minute when referring to rate of flow. 2. gallons per thousand cubic feet when referring to natural gas in terms of chromatograph analysis or theoretical gallons.

gravity *n*: 1. the attraction exerted by the earth's mass on objects at its surface. 2. the weight of a body. See *API gravity, relative density, specific gravity.*

H **hammer drill** *n*: a drilling tool that, when placed in the drill stem just above a roller cone bit, delivers high-frequency percussion blows to the rotating bit. Hammer drilling combines the basic features of rotary and cable-tool drilling (i.e., bit rotation and percussion).

hardfacing *n*: an extremely hard material, usually crushed tungsten carbide, that is applied to the outside surfaces of tool joints, drill collars, stabilizers, and other rotary drilling tools to minimize wear when they are in contact with the wall of the hole.

hazardous materials (HAZMAT) *n pl*: (DOT) substances or materials in quantities or forms that may pose an unreasonable risk to health, safety, or property when stored, transported, or used in commerce.

hhp *abbr*: hydraulic horsepower.

hold-down pressure *n*: hydrostatic pressure developed by the weight of the drilling fluid exerted on the bottom of the hole that tends to prevent cuttings from moving up the annulus.

horizontal drilling *n*: deviation of the borehole at least 80° from vertical so that the borehole penetrates a productive formation in a manner parallel to the formation. A single horizontal hole can effectively drain a reservoir and eliminate the need for several vertical boreholes.

horsepower *n*: a unit of measure of work done by a machine. One horsepower equals 33,000 foot-pounds per minute. (Kilowatts are used to measure power in the international, or SI, system of measurement.)

hydraulic *adj*: 1. of or relating to water or other liquid in motion. 2. operated, moved, or effected by water or liquid.

hydraulic horsepower *n*: a measure of the power of a fluid under pressure.

hydrostatic pressure *n*: the force exerted by a body of fluid at rest. It increases directly with the density and the depth of the fluid and is expressed in pounds per square inch or kilopascals. The hydrostatic pressure of fresh water is 0.433 pounds per square inch per foot (9.792 kilopascals per metre) of depth. In drilling, the term refers to the pressure exerted by the drilling fluid in the wellbore. In a water drive field, the term refers to the pressure that may furnish the primary energy for production.

I **impermeable** *adj*: preventing the passage of fluid. A formation may be porous yet impermeable if there is an absence of connecting passages between the voids within it. See *permeability.*

in. *abbr*: inch.

insert *n*: a cylindrical object, rounded, blunt, or chisel-shaped on one end and usually made of tungsten carbide, that is inserted in the cones of a bit, the cutters of a reamer, or the blades of a stabilizer to form the cutting element of the bit or the reamer or the wear surface of the stabilizer. Also called a compact.

insert bit *n*: see *tungsten carbide bit*.

invert-emulsion mud *n*: an oil mud in which fresh or salt water is the dispersed phase and diesel, crude, or some other oil is the continuous phase. See *oil mud*.

J

jet bit *n*: a drilling bit having replaceable nozzles through which the drilling fluid is directed in a high-velocity stream to the bottom of the hole to improve the efficiency of the bit.

junk *n*: metal debris lost in the hole. Junk may be a lost bit, pieces of a bit, milled pieces of pipe, wrenches, or any relatively small object that impedes drilling and must be fished out of the hole.

K

kelly *n*: the heavy steel tubular device, four- or six-sided, suspended from the swivel through the rotary table and connected to the top joint of drill pipe to turn the drill stem as the rotary table turns. It has a bored passageway that permits fluid to be circulated into the drill stem and up the annulus, or vice versa. Kellys manufactured to API specifications are available only in four- or six-sided versions, are either 40 or 54 feet (12 to 16 metres) long, and have diameters as small as 2½ inches (6 centimetres) and as large as 6 inches (15 centimetres).

keyseat *n*: an undergauge channel or groove cut in the side of the borehole and parallel to the axis of the hole. A keyseat results from the rotation of pipe on a sharp bend in the hole.

kick *n*: an entry of water, gas, oil, or other formation fluid into the wellbore during drilling. It occurs because the pressure exerted by the column of drilling fluid is not great enough to overcome the pressure exerted by the fluids in the formation drilled. If prompt action is not taken to control the kick, or kill the well, a blowout may occur.

L

lb *abbr*: pound.

lb/ft³ *abbr*: pounds per cubic foot.

lignins *n pl*: naturally occurring special lignites, e.g., leonardite, that are produced by strip mining from special lignite deposits. Used primarily as thinners and emulsifiers.

lignosulfonate *n*: an organic drilling fluid additive derived from by-products of a papermaking process using sulfite. It minimizes fluid loss and reduces mud viscosity.

lime *n*: a caustic solid that consists primarily of calcium oxide (CaO). Many forms of CaO are called lime, including the various chemical and physical forms of quicklime, hydrated lime, and even calcium carbonate. Limestone is sometimes called lime.

lime mud *n*: 1. a calcite-rich sediment that may give rise to shaly limestone. 2. a drilling mud that is treated with lime to provide a source of soluble calcium in the filtrate to obtain desirable mud properties for drilling in shale or clay formations.

limestone *n*: a sedimentary rock rich in calcium carbonate that sometimes serves as a reservoir rock for petroleum.

liquid *n*: a state of matter in which the shape of the given mass depends on the containing vessel, but the volume of the mass is independent of the vessel. A liquid is a fluid that is almost incompressible.

liquid phase *n*: in drilling fluids, that part of the fluid that is liquid. Normally, the liquid phase of a drilling fluid is water, oil, or a combination of water and oil.

litre (L) *n*: a unit of metric measure of capacity equal to the volume occupied by 1 kilogram of water at 4°C and at the standard atmospheric pressure of 760 millimetres.

logging while drilling (LWD) *n*: logging measurements obtained by measurement-while-drilling techniques as the well is being drilled.

lost circulation *n*: the quantities of whole mud lost to a formation, usually in cavernous, fissured, or coarsely permeable beds. Evidenced by the complete or partial failure of the mud to return to the surface as it is being circulated in the hole. Lost circulation can lead to a blowout and, in general, can reduce the efficiency of the drilling operation. Also called lost returns.

lost circulation additives *n pl*: materials added to the mud in varying amounts to control or prevent lost circulation. Classified as fiber, flake, or granular.

lost circulation material (LCM) *n*: a substance added to cement slurries or drilling mud to prevent the loss of cement or mud to the formation. See *lost circulation additives*.

low clay-solids mud *n*: a drilling mud that contains a minimum amount of solid materials that is used in rotary drilling when possible because it can provide fast drilling rates.

low-solids mud *n*: a drilling mud that contains a minimum amount of solid material and that is used in rotary drilling when possible because it can provide fast drilling rates.

LWD *abbr*: logging while drilling. Downhole electronic instruments transmit data to the surface. Rock type, density, and other information is displayed at the surface while drilling is in progress.

M

m *sym*: metre.

m² *abbr*: square metre.

m³ *abbr*: cubic metre.

make a trip *v*: to hoist the drill stem out of the wellbore to perform one of a number of operations, such as changing bits or taking a core, and so forth, and then to return the drill stem to the wellbore.

make hole *v*: to deepen the hole made by the bit, i.e., to drill ahead.

marl *n*: a semisolid or unconsolidated clay, silt, or sand.

matrix *n*: 1. in rock, the fine-grained material between larger grains in which the larger grains are embedded. A rock matrix may be composed of fine sediments, crystals, clay, or other substances. 2. the material in which the diamonds on a diamond bit are set.

measurement while drilling (MWD) *n*: 1. directional and other surveying during routine drilling operations to determine the angle and direction by which the wellbore deviates from the vertical. 2. any system of measuring downhole conditions during routine drilling operations.

megapascal *n*: one million pascals.

metre (m) *n*: the fundamental unit of length in the international system of measurement (SI). It is equal to about 3.28 feet, 39.37 inches, or 100 centimetres.

mist drilling *n*: a drilling technique that uses air or gas to which a foaming agent has been added. Also called foam drilling.

mix mud *v*: to prepare drilling fluids from a mixture of water or other liquids and any one or more of the various dry mud-making materials (such as clay, weighting materials, and chemicals).

mud *n*: the liquid circulated through the wellbore during rotary drilling and workover operations. Although it was originally a suspension of earth solids (especially clays) in water, the mud used in modern drilling operations is a more complex, three-phase mixture of liquids, reactive solids, and nonreactive solids. The liquid phase may be fresh water, diesel oil, or crude oil and may contain one or more conditioners. See *drilling fluid, drilling mud*.

mud additive *n*: any material added to drilling fluid to change some of its characteristics or properties.

mud column *n*: the borehole when it is filled or partially filled with drilling mud.

mud conditioning *n*: the treatment and control of drilling mud to ensure that it has the correct properties. Conditioning may include the use of additives, the removal of sand or other solids, the removal of gas, the addition of water, and other measures to prepare the mud for conditions encountered in a specific well.

mud engineer *n*: an employee of a drilling fluid supply company whose duty it is to test and maintain the drilling mud properties that are specified by the operator.

mud logging *n*: the recording of information derived from examination and analysis of formation cuttings made by the bit and of mud circulated out of the hole. A portion of the mud is diverted through a gas-detecting device. Cuttings brought up by the mud are examined under ultraviolet light to detect the presence of oil or gas. Mud logging is often carried out in a portable laboratory set up at the well.

mud program *n*: a plan or procedure, with respect to depth, for the type and properties of drilling fluid to be used in drilling a well. Some factors that influence the mud program are the casing program and such formation characteristics as type, competence, solubility, temperature, and pressure.

mud report *n*: a special form that is filled out by the mud engineer to record the properties of the drilling mud used while a well is being drilled.

mud solids *n pl*: the solid components of drilling mud. They may be added intentionally (e.g., barite), or they may be introduced into the mud from the formation as the bit drills ahead. The term is usually used to refer to the latter.

mud weight *n*: a measure of the density of a drilling fluid expressed as pounds per gallon, pounds per cubic foot, or kilograms per cubic metre. Mud weight is directly related to the amount of pressure the column of drilling mud exerts at the bottom of the hole.

MWD *abbr*: measurement while drilling. Sensors transmit data through an integrated rig-site computer system that allows monitoring and control of many drilling variables while drilling is in progress. Among the variables observed are pit volume, pump strokes, depth, hook load, torque, pressures, direction, and other critical data.

N

natural clays *n pl*: clays that are encountered when drilling various formations; they may or may not be incorporated purposely into the mud system.

natural mud *n*: a drilling fluid containing essentially clay and water; no special or expensive chemicals or conditioners are added. Also called conventional mud.

nonreactive phase *n*: that part of a liquid drilling mud that consists of solids or other chemicals that do not react with the liquid (or other chemicals in the liquid) part of the mud. Barite, for example, is nonreactive.

normal circulation *n*: the smooth, uninterrupted circulation of drilling fluid down the drill stem, out the bit, up the annular space between the pipe and the hole, and back to the surface. Compare *reverse circulation*.

nozzle *n*: a passageway through jet bits that causes the drilling fluid to be ejected from the bit at high velocity. The jets of mud clear the bottom of the hole. Nozzles come in different sizes that can be interchanged on the bit to adjust the velocity with which the mud exits the bit.

O

offset *n*: see *cone offset*.

oil-base mud *n*: a drilling or workover fluid in which oil is the continuous phase and which contains from less than 2 percent and up to 5 percent water. This water is spread out, or dispersed, in the oil as small droplets. See *invert emulsion mud, oil mud*.

oil-emulsion mud *n*: a water-base mud in which water is the continuous phase and oil is the dispersed phase. The oil is spread out, or dispersed, in the water in small droplets, which are tightly emulsified so that they do not settle out. Because of its lubricating abilities, an oil-emulsion mud increases the drilling rate and ensures better hole conditions than other muds. Compare *oil mud*.

oil mud *n*: a drilling mud, e.g., oil-base mud and invert-emulsion mud, in which oil is the continuous phase. It is useful in drilling certain formations that may be difficult or costly to drill with water-base mud. Compare *oil-emulsion mud*.

operator *n*: the person or company, either proprietor or lessee, actually operating an oilwell or lease, generally the oil company that engages the drilling contractor.

optimization *n*: the manner of planning and drilling a well so that the most usable hole will be drilled for the least money.

overbalanced drilling *n*: drilling in which the hydrostatic pressure of the mud column exceeds formation pressure.

overburden *n*: the strata of rock that overlie the stratum of interest in drilling.

pcf *abbr*: pounds per cubic foot.

PDC *abbr*: polycrystalline diamond compact.

PDC bit *n*: a special type of diamond drilling bit that does not use roller cones. Instead, polycrystalline diamond inserts are embedded into a matrix on the bit. PDC bits are often used to drill very hard, abrasive formations, but also find use in drilling medium and soft formations.

permeability *n*: 1. a measure of the ease with which a fluid flows through the connecting pore spaces of rock or cement. The unit of measurement is the millidarcy. 2. fluid conductivity of a porous medium. 3. ability of a fluid to flow within the interconnected pore network of a porous medium. See *absolute permeability*, *effective permeability*, *relative permeability*.

pH *abbr*: an indicator of the acidity or alkalinity of a substance or solution, represented on a scale of 0–14, 0–6.9 being acidic, 7 being neither acidic nor basic (i.e., neutral), and 7.1–14 being basic. These values are based on hydrogen ion content and activity.

phase *n*: a portion of a physical system that is liquid, gas, or solid, that is homogeneous throughout, that has definite boundaries, and that can be separated from other phases. The three phases of drilling mud are the liquid or continuous phase, the reactive or colloidal phase, and the nonreactive phase.

pH control agent *n*: a chemical added to the drilling fluid to control or increase the pH of the mud. Normally, the mud should have a pH higher than seven so that it is alkaline.

pit level *n*: height of drilling mud in the mud tanks, or pits.

pit-level indicator *n*: one of a series of devices that continuously monitor the level of the drilling mud in the mud tanks. The indicator usually consists of float devices in the mud tanks that sense the mud level and transmit data to a recording and alarm device (a pit-volume recorder) mounted near the driller's position on the rig floor. If the mud level drops too low or rises too high, the alarm sounds to warn the driller of lost circulation or a kick.

polymer *n*: a substance that consists of large molecules formed from smaller molecules in repeating structural units (monomers). In oilfield operations, various types of polymers are used to thicken drilling mud, fracturing fluid, acid, water, and other liquids. See *polymer mud*.

polymer mud *n*: a drilling mud to which a polymer has been added to increase the viscosity of the mud.

pore *n*: an opening or space within a rock or mass of rocks, usually small and often filled with some fluid (water, oil, gas, or all three). Compare *vug*.

porosity *n*: 1. the condition of being porous (such as a rock formation). 2. the ratio of the volume of empty space to the volume of solid rock in a formation, indicating how much fluid a rock can hold. See *absolute porosity*, *effective porosity*, *pore*.

porous *adj*: having pores, or tiny openings, as in rock.

pounds per cubic foot *n*: a measure of the density of a substance (such as drilling fluid).

pounds per gallon (ppg) *n*: a measure of the density of a fluid (such as drilling mud).

P

ppg *abbr*: pounds per gallon.

pressure *n*: the force that a fluid (liquid or gas) exerts uniformly in all directions within a vessel, pipe, hole in the ground, and so forth, such as that exerted against the inner wall of a tank or that exerted on the bottom of the wellbore by a fluid. Pressure is expressed in terms of force exerted per unit of area, as pounds per square inch, or in kilopascals.

pressure gradient *n*: 1. a scale of pressure differences in which there is a uniform variation of pressure from point to point. For example, the pressure gradient of a column of water is about 0.433 pounds per square inch per foot (9.794 kilopascals per metre) of vertical elevation. The normal pressure gradient in a formation is equivalent to the pressure exerted at any given depth by a column of 10 percent salt water extending from that depth to the surface (0.465 pounds per square inch per foot, or 10.518 kilopascals per metre). 2. the change (along a horizontal distance) in atmospheric pressure. Isobars drawn on weather maps display the pressure gradient.

pressure loss *n*: the drilling fluid's loss of hydraulic pressure after it leaves the pump. Some pressure is lost due to friction, but the main loss occurs when the fluid leaves the bit nozzles. See *friction loss*.

pressure surge *n*: a sudden and usually short-duration increase in pressure. When pipe or casing is run into a hole too rapidly, an increase in the hydrostatic pressure results, which may be great enough to create lost circulation.

psi *abbr*: pounds per square inch.

pump *n*: a device that increases the pressure on a fluid. Types of pumps used in the circulating system include the mud pump (a reciprocating pump), centrifugal pumps, and the downhole pump.

pump liner *n*: a cylindrical, accurately machined, metallic section that forms the working barrel of some reciprocating pumps. Liners are an inexpensive means of replacing worn cylinder surfaces, and in some pumps they provide a method of conveniently changing the displacement and capacity of the pumps.

pump stroke indicator *n*: an instrument that measures pump speed by counting the number of strokes per minute. Also called pump stroke counter.

R **rate of penetration (ROP)** *n*: the speed with which the bit drills the formation.

relative density *n*: 1. the ratio of the weight of a given volume of a substance at a given temperature to the weight of an equal volume of a standard substance at the same temperature. For example, if 1 cubic inch of water at 39°F (3.9°C) weighs 1 unit and 1 cubic inch of another solid or liquid at 39°F weighs 0.95 unit, then the relative density of the substance is 0.95. In determining the relative density of gases, the comparison is made with the standard of air or hydrogen. 2. the ratio of the mass of a given volume of a substance to the mass of a like volume of a standard substance, such as water or air.

relative permeability *n*: the ratio of effective permeability to absolute permeability. The relative permeability of rock to a single fluid is 1.0 when only that fluid is present, and 0.0 when the presence of another fluid prevents all flow of the given fluid. Compare *absolute permeability, effective permeability*.

reverse circulation *n*: the course of drilling fluid downward through the annulus and upward through the drill stem, in contrast to normal circulation in which the course is downward through the drill stem and upward through the annulus. Seldom used in open hole, but frequently used in workover operations. Also referred to as "circulating the short way," since returns from bottom can be obtained more quickly than in normal circulation. Compare *normal circulation*.

rock bit *n*: see *roller cone bit*.

roller cone bit *n*: a drilling bit made of two, three, or four cones, or cutters, that are mounted on extremely rugged bearings. The surface of each cone is made of rows of steel teeth or rows of tungsten carbide inserts. Also called rock bit.

ROP *abbr*: rate of penetration.

rotary *n*: the machine used to impart rotational power to the drill stem while permitting vertical movement of the pipe for rotary drilling. Modern rotary machines have a special component, the rotary or master bushing, to turn the kelly bushing, which permits vertical movement of the kelly while the stem is turning.

rotary hose *n*: a steel-reinforced, flexible hose that is installed between the standpipe and the swivel or top drive. It conducts drilling mud from the standpipe to the swivel or top drive. Also called the kelly hose or the mud hose.

rotary speed *n*: the speed, measured in revolutions per minute, at which the rotary table is operated.

rotary table *n*: the principal piece of equipment in the rotary table assembly; a turning device used to impart rotational power to the drill stem while permitting vertical movement of the pipe for rotary drilling. The master bushing fits inside the opening of the rotary table; it turns the kelly bushing, which permits vertical movement of the kelly while the stem is turning.

rotating head *n*: a sealing device used to close off the annular space around the kelly in drilling with pressure at the surface, usually installed above the main blowout preventers. A rotating head makes it possible to drill ahead even when there is pressure in the annulus that the weight of the drilling fluid is not overcoming; the head prevents the well from blowing out. It is used mainly in the drilling of formations that have low permeability. The rate of penetration through such formations is usually rapid.

rotor *n*: 1. a device with vanelike blades attached to a shaft. The device turns or rotates when the vanes are struck by a fluid directed there by a stator. 2. the rotating part of an induction-type alternating current electric motor.

rpm *abbr*: revolutions per minute.

S

sandstone *n*: a sedimentary rock composed of individual mineral grains of rock fragments between 0.06 and 2 millimetres (0.002 and 0.078 inches) in diameter and cemented together by silica, calcite, iron oxide, and so forth. Sandstone is commonly porous and permeable and therefore a likely type of rock in which to find a petroleum reservoir.

shale *n*: a fine-grained sedimentary rock composed mostly of consolidated clay or mud. Shale is the most frequently occurring sedimentary rock.

shear *n*: action or stress that results from applied forces and that causes or tends to cause two adjoining portions of a substance or body to slide relative to each other in a direction parallel to their plane of contact.

silicon tetrafluoride *n*: a gas that can be readily absorbed by water and that is used to seal off water-bearing formations in air drilling.

slack off *v*: to lower a load or ease up on a line. A driller will slack off on the brake to put additional weight on the bit.

sloughing (pronounced "sluffing") *n*: see *caving*.

sodium acid pyrophosphate (SAPP) *n*: a thinner used in combination with barite, caustic soda, and fresh water to form a plug and seal off a zone of lost circulation.

sodium carbonate *n*: Na_2CO_3, used extensively for treating various types of calcium contamination. Also called soda ash.

sodium hydroxide *n*: see *caustic soda*.

specific gravity *n*: the ratio of the density, or weight, of a substance to the density of a reference substance. For liquids and solids, the reference substance is water.

spot *v*: to pump a designated quantity of a substance (such as acid or cement) into a specific interval in the well. For example, 10 barrels (1,590 litres) of diesel oil may be spotted around an area in the hole in which drill collars are stuck against the wall of the hole in an effort to free the collars.

spurt loss *n*: the initial loss of drilling mud solids by filtration, making formations easier to drill. See *filtration loss, surge loss*.

sq *abbr*: square.

standpipe *n*: a vertical pipe rising along the side of the derrick or mast, which joins the discharge line leading from the mud pump to the rotary hose and through which mud is pumped into the hole.

stator *n*: the long helical shaft in a downhole motor that rotates as mud is forced down through the power section of the motor. Also referred to as a rotor.

steel-tooth bit *n*: a roller cone bit in which the surface of each cone is made up of rows of steel teeth. Also called a milled-tooth bit or milled bit.

strata *n pl*: distinct, usually parallel, and originally horizontal beds of rock. An individual bed is a stratum.

stringer *n*: 1. an extra support placed under the middle of racked pipe to keep the pipe from sagging. 2. a relatively narrow splinter of a rock formation that is stratigraphically disjointed, interrupts the consistency of another formation, and makes drilling that formation less predictable. A shale formation, for example, may be broken by a stringer of sandstone.

surfactant *n*: a soluble compound that concentrates on the surface boundary between two substances such as oil and water and reduces the surface tension between the substances. The use of surfactants permits the thorough surface contact or mixing of substances that ordinarily remain separate. Surfactants are used in the petroleum industry as additives to drilling mud and to water during chemical flooding.

surge loss *n*: the flux of fluids and solids that occurs in the initial stages of any filtration before pore openings are bridged and a filter cake is formed. Also called spurt loss.

suspension *n*: a mixture of small nonsettling particles of solid material within a gaseous or liquid medium.

swivel *n*: a rotary tool that is hung from the hook and the traveling block to suspend and permit free rotation of the drill stem. It also provides a connection for the rotary hose and a passageway for the flow of drilling fluid into the drill stem.

T

thinning agent *n*: a special chemical or combination of chemicals that, when added to a drilling mud, reduces its viscosity.

top drive *n*: a device similar to a power swivel that is used in place of the rotary table to turn the drill stem. It also includes power tongs. Modern top drives combine the elevator, the tongs, the swivel, and the hook. Even though the rotary table assembly is not used to rotate the drill stem and bit, the top-drive system retains it to provide a place to set the slips to suspend the drill stem when drilling stops.

torque *n*: the turning force that is applied to a shaft or other rotary mechanism to cause it to rotate or tend to do so. Torque is measured in foot-pounds, joules, newton-metres, and so forth.

toxicity *n*: the ability of a substance to be poisonous if inhaled, swallowed, absorbed, or introduced into the body through cuts or breaks in the skin.

traveling block *n*: an arrangement of pulleys, or sheaves, through which drilling line is reeved and which moves up and down in the derrick or mast. See *block*.

tricone bit *n*: a type of bit in which three cone-shaped cutting devices are mounted in such a way that they intermesh and rotate together as the bit drills. The bit body may be fitted with nozzles, or jets, through which the drilling fluid is discharged.

tripping *n*: the operation of hoisting the drill stem out of and returning it to the wellbore. See *make a trip*.

trip tank indicator *n*: a device installed on a trip tank that shows the amount of mud being removed from or added to the trip tank.

tungsten carbide bit *n*: a type of roller cone bit with inserts made of tungsten carbide. Also called tungsten carbide insert bit.

twistoff *n*: a complete break in pipe caused by metal fatigue.

twist off *v*: to break something in two or to break apart, such as the head of a bolt or the drill stem.

U

unconsolidated formation *n*: a loosely arranged, apparently unstratified section of rock.

underbalanced drilling *v*: to carry on drilling operations with a mud whose density is such that it exerts less pressure on bottom than the pressure in the formation while maintaining a seal (usually with a rotating head) to prevent the well fluids from blowing out under the rig. Drilling under pressure is advantageous in that the rate of penetration is relatively fast; however, the technique requires extreme caution.

undergauge hole *n*: that portion of a borehole drilled with an undergauge bit.

V **viscosity** *n*: a measure of the resistance of a fluid to flow. Resistance is brought about by the internal friction resulting from the combined effects of cohesion and adhesion. The viscosity of petroleum products is commonly expressed in terms of the time required for a specific volume of the liquid to flow through a capillary tube of a specific size at a given temperature.

viscous *adj*: having a high resistance to flow.

vug *n*: 1. a cavity in a rock. 2. a small cavern, larger than a pore but too small to contain a person. Typically found in limestone subject to groundwater leaching.

W **wall-building ability** *n*: the ability of a drilling mud to plaster the wall of the hole with solids from the drilling mud.

wall cake *n*: also called filter cake or mud cake. See *filter cake*.

water-base mud *n*: a drilling mud in which the continuous phase is water. In water-base muds, any additives are dispersed in the water. Compare *oil-base mud*.

watercourse *n*: a hole inside a bit through which drilling fluid from the drill stem is directed.

weighting material *n*: a material that has a high specific gravity and is used to increase the density of drilling fluids or cement slurries.

wellbore *n*: a borehole; the hole drilled by the bit. A wellbore may have casing in it or it may be open (uncased); or part of it may be cased, and part of it may be open. Also called a borehole or hole.

wildcat *n*: 1. a well drilled in an area where no oil or gas production exists. 2. (nautical) the geared sheave of a windlass used to pull anchor chain. *v*: to drill wildcat wells.

Review Questions
LESSONS IN ROTARY DRILLING
Unit II, Lesson 1: Making Hole

Multiple Choice

Pick the *best* answer from the choices and place the letter of that answer in the blank provided.

_____ 1. You are drilling a deep well in West Texas. The wellbore will likely encounter—
- a. shallow tar sands.
- b. overpressured gas zones.
- c. petrified wood.
- d. gumbo.

_____ 2. On a particular well, the bit cost $5,200 and, while rotating on bottom for 74 hours, it drilled 615 ft. The crew then pulled the bit and ran a new one back to bottom in 8.5 hours. The rig costs are $325 per hour. Therefore, the cost per ft (m) to drill this portion of the hole is about—
- a. $48 ($154).
- b. $52 ($171).
- c. $74 ($243).
- d. $104 ($341).

_____ 3. Optimized drilling is—
- a. drilling a well safely, no matter what it costs.
- b. drilling a well as cheaply as possible without regard to any factor besides costs.
- c. drilling a well as quickly as possible, because the faster the well is drilled, the cheaper it is.
- d. drilling a well as cheaply as possible while taking into account the safety of people, property, and the environment.

_____ 4. Good well planning can reduce costs; however—
- a. the last wells drilled in an area usually cost less than the first wells.
- b. the first wells drilled in an area usually cost less than the last wells.
- c. the first and last wells drilled in an area usually cost about the same.
- d. costs are not important when it comes to drilling a well.

_____ 5. When selecting a rig to drill a particular well, factors include—
- a. depth rating.
- b. hoisting capacity.
- c. pump size and how many.
- d. all of the above

True or False

Put a T for *true* or an F for *false* in the blank next to each statement.

_____ 6. In general, the most important consideration affecting bit design is the type of rock the bit will be drilling.

_____ 7. The only two types of bit are rock bits and roller cone bits.

_____ 8. Three materials used to manufacture bits are steel, tungsten carbide alloy, and diamonds, either natural or synthetic.

_____ 9. A bit designed to drill soft formations usually does not have much offset.

_____ 10. In general, steel-tooth bits designed to drill hard formations have closely spaced, short teeth.

_____ 11. Generally, insert bits are less expensive than steel-tooth bits.

_____ 12. Fixed-head bits have no moving parts.

_____ 13. The synthetic diamonds used in PDC bits are more stable than natural diamonds.

_____ 14. Thermally stable polycrystalline (TSP) bits are less stable than PDCs.

_____ 15. A hybrid bit combines a roller-cone bit with a tungsten carbide insert bit.

Multiple Choice

Pick the *best* answer from the choices and place the letter of that answer in the blank provided.

_____ 16. Common reasons for pulling a bit include—
a. increase in torque.
b. junk in the hole.
c. decrease in penetration rate.
d. all of the above

_____ 17. Dull roller cone bits are graded on the basis of—
a. cutter wear.
b. bearing wear.
c. gauge wear.
d. all of the above

_____ 18. Causes of broken inserts include—
a. inserts too long for hard formation drilling.
b. junk in the hole.
c. excessive weight on bit.
d. all of the above

_____ 19. Drilling performance records include—
  a. bit records.
  b. daily drilling reports.
  c. both a and b
  c. neither a nor b

_____ 20. If the right bit has been selected and bit hydraulics are good, the driller can usually increase the rate of penetration by—
  a. decreasing weight on bit and increasing pump speed.
  b. increasing weight on bit and rotary speed (rpm).
  c. decreasing weight on bit and rotary speed (rpm).
  d. increasing weight on bit and decreasing rotary speed (rpm).

Fill in the Blanks

Fill in the blanks with an *appropriate* word or phrase. Pick the correct term from those listed below.

weight on bit	failure
bent sub	bit life
decreased	pump capacity
compressive strength of rock	breakage
friction	number of reusable bits

In general, as weight on bit is increased, rotary speed (rpm) should be 21._____. Soft formation bits are usually run with lower 22. _____ ___ _____ than hard formation bits. In hard formations, bit weights are usually heavier than in soft formations so that the action of the bit can overcome 23. _____ _____ __ _____. At high rotary speeds, drill string 24. _____ is more common than at low rotary speeds. Maintaining a steady weight on the bit enhances penetration rates and increases both 25. _____ _____ _____ and 26. _____ ___ _____ _____. Slanted holes and holes with sharp doglegs cause increased rotary torque because of increased 27. _____ between the drill string and walls of the hole. High rpm, torque, and weight on the drill stem can result in joint failure, twistoffs, and tooth 28. _____. Downhole motors require more 29. _____ _____ than regular rotary drilling. The turn in horizontal drilling is initiated by using a 30. _____ _____ above the motor.

True or False

Put a T for *true* or an F for *false* in the blank next to each statement.

_____ 31. As more of the drill string weight is suspended by the crown block less weight is applied to the bit.

_____ 32. Drilling mud is made up of three types of material, a liquid and two solids.

_____ 33. One type of reactive solid is barite.

_____ 34. Mud weight (density) has little effect on the penetration rate.

_____ 35. Viscosity is a measure of a mud's resistance to flow.

_____ 36. Wall cake is a build up of the solids in the mud on the wall of the hole.

_____ 37. fluid loss is unrelated to wall-cake thickness.

_____ 38. Oil-base muds are seldom used because they do not perform as well as water-base muds.

_____ 39. In general, using air or gas as a drilling fluid results in lower rates of penetration.

_____ 40. Surfactants (foaming agents) can sometimes be added to air or gas to overcome problems associated with formations producing water.

Multiple Choice

Pick the *best* answer from the choices and place the letter of that answer in the blank provided.

_____ 41. Bit hydraulic horsepower—
    a. is not important to rate of penetration.
    b. is a measure drilling mud density.
    c. is a measure of a drilling fluid's ability to clean the bottom of the hole.
    d. decreases the rate of penetration if air or gas is used.

_____ 42. Hydraulic horsepower is determined by—
    a. pump output and circulating pressure.
    b. pump output only.
    c. circulating pressure only.
    d. weight on the bit.

_____ 43. The largest amount of pump pressure lost because of circulation should be—
    a. in the surface equipment and standpipe.
    b. in the drill string.
    c. at the bit.
    d. in the annulus.

_____ 44. One way in which bit hydraulic horsepower is controlled is by—
   a. the size of the bit nozzles.
   b. using a PDC instead of a roller-cone bit.
   c. increasing the size of the rig's prime movers (engines).
   d. the bit manufacturer.

_____ 45. A large influence on the drilling rate is the—
   a. type of formation.
   b. size of the drill string.
   c. size of the bit.
   d. none of the above

_____ 46. A drilling break occurs when—
   a. the bit cones disintegrate.
   b. the driller increases weight on the bit.
   c. the rate of penetration increases.
   d. rotary speed decreases.

_____ 47. Because rock strength usually increases with depth, the driller usually has to—
   a. increase the weight on the bit.
   b. decrease the weight on the bit.
   c. switch to a softer formation.
   d. pick up off bottom and shut the well in.

_____ 48. Some formations react to water-base muds by—
   a. becoming harder.
   b. turning into shale.
   c. swelling and sloughing.
   d. changing the water-base mud into an oil-base mud.

_____ 49. Formation dip, a formation's inclination from horizontal, can cause a hole to—
   a. deviate from vertical.
   b. straighten out.
   c. cave in.
   d. none of the above

_____ 50. Underbalanced drilling is—
   a. drilling with hydrostatic pressure higher than formation pressure.
   b. drilling with hydrostatic pressure lower than formation pressure.
   c. drilling with the rig derrick leaning to one side.
   d. drilling with high mud weight to overcome formation pressure.

Answers to Review Questions
LESSONS IN ROTARY DRILLING
Unit II, Lesson 1: Making Hole

Multiple Choice

1. b
2. b
3. d
4. a
5. d

True or False

6. T
7. F
8. T
9. F
10. T
11. F
12. T
13. F
14. F
15. F

Multiple Choice

16. d
17. d
18. d
19. c
20. b

Fill in the Blanks

21. decreased
22. weight on bit
23. compressive strength of rock
24. failure

25. bit life
26. number of reusable bits
27. friction
28. breakage
29. pump capacity
30. bent sub

True or False

31. T
32. T
33. F
34. F
35. T
36. T
37. F
38. F
39. F
40. T

Multiple Choice

41. c
42. a
43. c
44. a
45. b
46. c
47. a
48. c
49. a
50. b